Construction Science

Volume 1

Construction Science

Volume 1

BRIAN W. BOUGHTON
C.Eng., M.I.Struct.E.
Advisory Officer for
Building, Civil Engineering and Surveying,
Business and Technician Education Council
and
DENIS GOODFELLOW
B.Sc., Cert.Ed. (Tech.)
Senior Lecturer in Science,
Hertfordshire College of Building

GRANADA
London Toronto Sydney New York

Granada Technical Books
Granada Publishing Ltd
8 Grafton Street, London W1X 3LA

First published in Great Britain by
Granada Publishing 1984

British Library Cataloguing in Publication Data

Boughton, Brian W.
Construction science.—(Granada construction TEC series)
Vol. 1
1. Building
I. Title II. Goodfellow, Denis
690 TH145

ISBN 0-246-12127-0

Typeset by Cambrian Typesetters, Aldershot, Hants
Printed and bound in Great Britain by
Richard Clay (The Chaucer Press) Ltd, Bungay, Suffolk

Contents

This book is intended to cover the content of TEC units:

Physical Science I
Construction Science II and
Materials in Construction II

together with elements of the units:

Construction Drawing I
Construction Processes I and
Construction Technology II

The material will be dealt with in an integrated manner, showing the relationship between the various units of study, and each chapter will include some assignment work. This is to encourage the reader to research outside the text to consolidate the material contained therein. Later chapters also contain exercises to practise the more mathematically orientated material.

The prime aim of the book is to develop the technician approach in applying scientific principles and techniques to construction practice: to promote logical thinking, including selection of materials, rather than to 'feed' information on how to do things by 'rote' learning. This is further enhanced by the inclusion of two projects at the end of the book.

Following the introduction (chapter one) the book is divided into three sections: I Materials, II Mechanics, III Environment. The text is related primarily to the forms of construction covered in the two volumes of Building and Civil Engineering Construction.

I Materials

1 Introduction

To many people the mention of the word 'science' produces an immediate reaction of bewilderment. It becomes a closed subject because of its treatment through school, where there is often a separation of the 'arts' and the 'sciences', and the general impression develops that it is very involved. This is because of the prominence of mathematical content or the language used. In all fairness this is a view that is often encouraged by engineers and mathematicians and has been further exaggerated by practitioners of the 'pseudo-sciences'. If they wish to confuse (or impress) somebody, or disrupt their logic in an argument, the introduction of technical jargon or mathematical equations might 'blind their listeners with science.' Unfortunately, this attitude often persists in the area of research, where papers of scientific merit, aimed at recognition towards a philosophy doctorate, are couched in language that expands the number of words to four times the amount required in an effort to impress.

The science covered in this book, and Volume 2, is based on that applied in the *real world* of construction and is couched in reasonably plain English that can be readily understood. The content is also based on fact, rather than theory, and attempts to show the logical reasons for the behaviour of materials, machines and buildings. There is a scientific reason, often linked to plain common sense, for the behaviour of anything connected with building and civil engineering construction and it is of critical importance that such reasoning is firmly established so that carelessness and wrong assumptions are eliminated. Although the text is separated into three parts, materials, mechanics and environment, it should be realised that all three are *interdependent* and cannot be viewed in isolation. By the same token, science, mathematics and technology are also interdependent and should be cross-related to develop a full understanding of constructional behaviour, function and efficiency.

Part I deals with materials, since these are the most important factor of any type of construction. It is not possible to construct anything without using materials, whether they are natural or fabricated, and a sound knowledge of their structure, behaviour and performance is necessary if they are to be dealt with sympathetically and used effectively.

Part II deals with mechanics, not in the way that a mechanical engineer thinks of the subject (relating it primarily to machinery), but as a means of developing an understanding of the physical and structural behaviour of materials, components and systems. This enables materials to be used effectively and economically to produce a cost-effective form of construction both initially and in the long term.

Part III deals with the environment: that existing externally, posing restrictions, problems and hazards; and that required internally (if an enclosed structure is to be formed), related to function, comfort and performance. The external environment already exists and is of such a variable nature that no two construction projects can be thought of as identical. The internal environment, however, is often of a form requiring very similar conditions, which means that the selection of materials and their utilisation must be treated on an individual basis. It is this variation of balance between the external and internal environment that separates construction from other forms of engineering and encourages the 'made to measure' approach rather than the 'off the peg' design used, for example, in the automobile industry.

Mechanics and environment are developed more fully in Volume 2, but materials are dealt with exclusively in Volume 1. This is because we are concerned with the basic physical and chemical properties of materials encountered in everyday forms of construction. Although it is of great interest to materials technologists to investigate the crystalline structure of a material, it is of more importance to a building or civil engineering technician to develop an appreciation of its workability, durability, functional effectiveness in a variety of situations and any characteristics that might limit its useful application for a given set of circumstances. It is also important to know of its 'tried and tested' performance based on *British Standards Specification* or *Agrément certification.* A knowledge of behaviour over a given temperature range is also valuable so that unwelcome stresses can be avoided. The material's durability in adverse conditions such as fire and damp is of obvious importance and its structural behaviour, if forming part of the load-bearing system, must fall within guaranteed limits.

The fundamental principles of applied and structural mechanics are introduced in Volume 1 so that they can be developed and consolidated in Volume 2. The treatment is to establish first principles that can then be applied rather than to teach masses of standard formulae, and to approach problems from a variety of directions to develop a thorough understanding of mechanical behaviour and performance based on logical thinking. In this area there is an obvious link with mathematics but this is seen as a necessary tool rather a discipline in its own right. The emphasis is always on the *practical* application to the types of system used in low-rise construction, earthworks or temporary works. This means that there is an equally valuable link with construction, this being of prime importance. Environmental science, in Volume 1, concentrates on three areas of study: thermal behaviour, electricity and sound. Since the energy crisis of the 1970s, thermal performance has assumed a greater importance than in earlier years, which is of obvious benefit to the user, particularly in domestic construction, and this cannot be divorced from the materials used. Also the problems of moisture movement, created by airborne and fabric-borne condensation contained within this study area, are related entirely to thermal performance. Electricity is used as both a temporary and a permanent supply and a working knowledge of its behaviour is of critical importance. Sound (its measurement, and the protection that can be offered to the public at large – or the occupants of the building) has in the past been something of the 'poor relation'. However, with the increase in traffic noise, the development of high-density domestic and commercial construction and the required functional performance of theatres, lecture halls and recording studios, an understanding of the subject can be seen as necessary to the comfort level and functional performance of construction projects and indeed to health and safety requirements.

One constructional problem that does not fit happily into any of the three areas of study, since it encompasses materials, mechanics and environment, is that of dampness in buildings. Because of this, it can best be dealt with in this chapter and may help to introduce some of the scientific concepts developed later in the book.

Dampness in Buildings

Dampness in buildings can be attributed to two main sources:

(a) *the external environment:* through rain-water, damp air, surface water, ground-water and leakage from supply or disposal services;

(b) *the internal environment:* through condensation from damp warm air, leakage of supply or disposal services and spillage in use of water.

Additionally, contributory factors that assist or exaggerate dampness from these sources are:

(a) poor selection of materials for specific locations;

(b) poor detailing of critical locations that are most subject to penetration;

(c) inadequate allowance for thermal or structural movement that can be expected to occur;

(d) structural deformation of parts of the building, which can produce conditions that allow the passage of moisture;

(e) inadequate supervision, during the construction process, to ensure that specification, detail and good practice are met;

(f) an inadequate maintenance programme causing deterioration of the fabric or structure;

(g) abuse of the building internally or externally.

Let us look first at the exterior to see the sources of damp and how they are countered.

External sources, effects and protective measures

The most obvious source of dampness is from rainfall. This, coupled with strong winds, can produce driving rain, which will penetrate gaps. It will also bounce off hard surfaces and stream down the face of the building seeking out weak points in the fabric.

Although it would seem logical that rain travels downwards, since it is falling, the nature of the fabric can cause it to move sideways or upwards because of capillary attraction. This can be seen in a simple experiment using glass tubes of various bore diameters. Place these in a bowl of water, keeping each tube vertical, and you will notice that the water will be attracted to the side walls of the tubes. On the biggest bore tube the curve of the water, as it climbs the side walls, can be seen clearly; and it will be equally noticeable that as the bores decrease so the water level climbs (see Fig. 1.1). This is caused by the action of *capillarity.*

The external face of the building can create conditions that promote capillarity in three obvious ways:

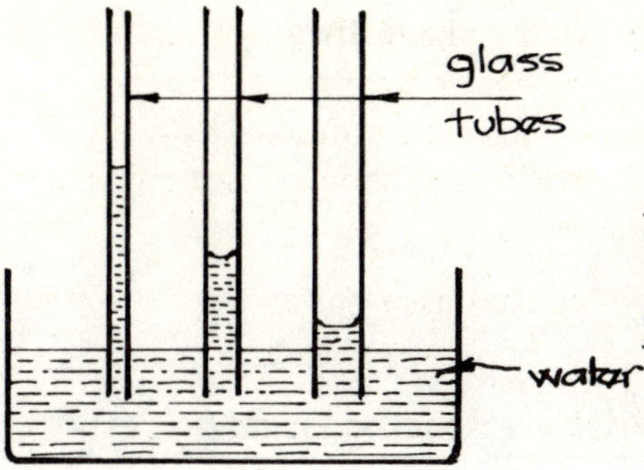

Fig 1.1 Variation of capillary movement

(a) fine joints at changes of material, such as mortar joints in brickwork, window and door frames abutting wall openings (and opening members within the components), intersections between wall/roof joints and lapping of components;

(b) fine surface cracks in renders, caused by shrinkage during their setting action;

(c) porosity of the materials used for the external faces of the building: the cellular structure of some materials provides an interconnection between pores so that water can travel through the material whereas others are either so densely packed that they do not contain air pockets or have voids that are completely separated.

Figure 1.2 shows stylised examples of these three routes of capillary attraction.

How then can these capillary paths be avoided, or at least stopped, to prevent water from reaching the inside of the building? The path must be broken by introducing a much wider gap, known as a capillary groove (though it is really *anti*-capillary); or by being physically separated from the internal fabric by a cavity; or barred from further progress by an impermeable barrier (such as a flashing, damp-proof course or sheathing). It can also help if the surface is 'layered' so that any gaps are staggered.

The following are some examples of how component manufacturers reduce the risk of capillarity occurring.

Plain tiles for roofing and wall cladding: These are curved slightly in cross-section so that each tile course is in contact with the next for only a short distance.

Timber sub-frames for metal windows: A weathering groove is formed around the external perimeter, both to break the capillary path and to allow for a weather strip or d.p.c. The internal perimeter

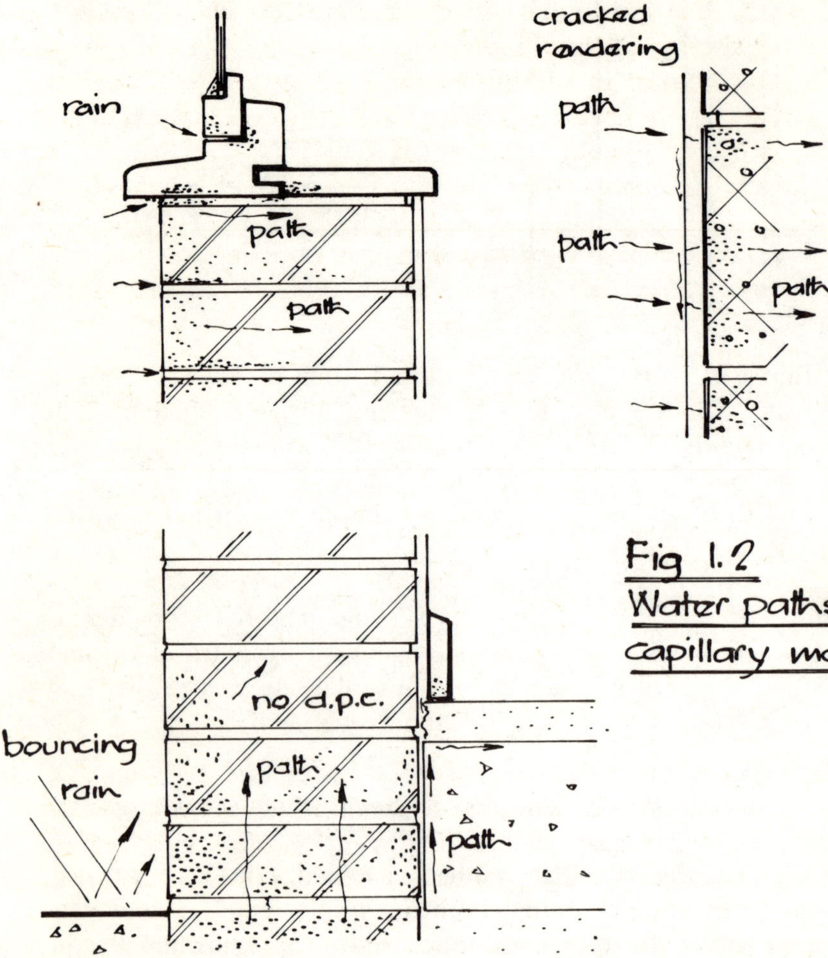

Fig 1.2
Water paths by capillary movement

is stepped to break the vertical and horizontal joints. Finally, the projecting cill member has a recess located forwards of the main vertical surface of the wall to stop water 'tracking back' to the point of intersection.

Metal window frames: These have a 'profiled' section to reduce the length of contact with the wall opening or sub-frame. Some have a Neoprene weather strip to stop the passage of water between the frame and the opening members. Most have a weather bar at the top of opening sections to prevent water lodging at the intersection.

Timber window frames: These have the external perimeter groove and cill recess as in the sub-frames. They are also stepped at opening sections, which additionally have capillary grooves. Weather bars are often fitted at the top of the window.

Timber door frames: In addition to the features identified on the window frames, these have a water bar fitted to the threshold (cill) and a weather bar on the outward opening door.

Shiplap cladding boards: Whether of timber, u.P.V.C., or plastic-coated aluminium, these are profiled to reduce the chance of capillary paths occurring and to form a distinct gap at the joints.

Feather-edged boarding: Because of its cross-section and lap, feather-edged boarding produces a minimum of contact area.

Sketches of these components are shown in Fig. 1.3. This list of components is by no means exhaustive and a study of manufacturers' catalogues is to be recommended.

Constructional detail can also help to reduce the risk of capillarity, the use of cavities and barriers being particularly effective. Let us consider the traditional brick wall. Until quite recently, as late as the mid-1950s, solid 'one brick' walling was a common feature of house building, producing a visually pleasant bonding pattern. Its resistance to water penetration from capillary action depended, however, on the density or porosity of the bricks used. It also posed other problems of interstitial condensation, more of which later in the chapter (and in chapter 6), which could result in damp from the outside meeting damp from the inside of the building! By changing the form of construction to cavity walling, this has been avoided and several other benefits have evolved. Precautions must be taken in forming the cavity and these are dealt with in *Building and Civil Engineering Construction* Volume 1. The most obvious of these is that where the cavity stops, a d.p.c. must be introduced – as an example, around window and door openings.

Alternatively, if a solid wall of lightweight concrete blocks is built, or timber-framed housing is constructed, the surface may be clad and a cavity formed between the cladding and the wall surface using battens. In such situations an additional precaution is to provide a 'breather paper' to the wall surface, which will not let water through but will allow vapour to pass out. The battens help to keep this 'paper' in place in addition to providing the fixing for the cladding, which may be tile hanging, shiplap boarding, vertical boarding, or additionally (in the case of timber-framed housing) cement/sand render on expanded metal lathing. In addition to forming a space across which the water will not travel, the cavity also allows ventilation, which assists in the drying of the fabric. Ideally, the width of

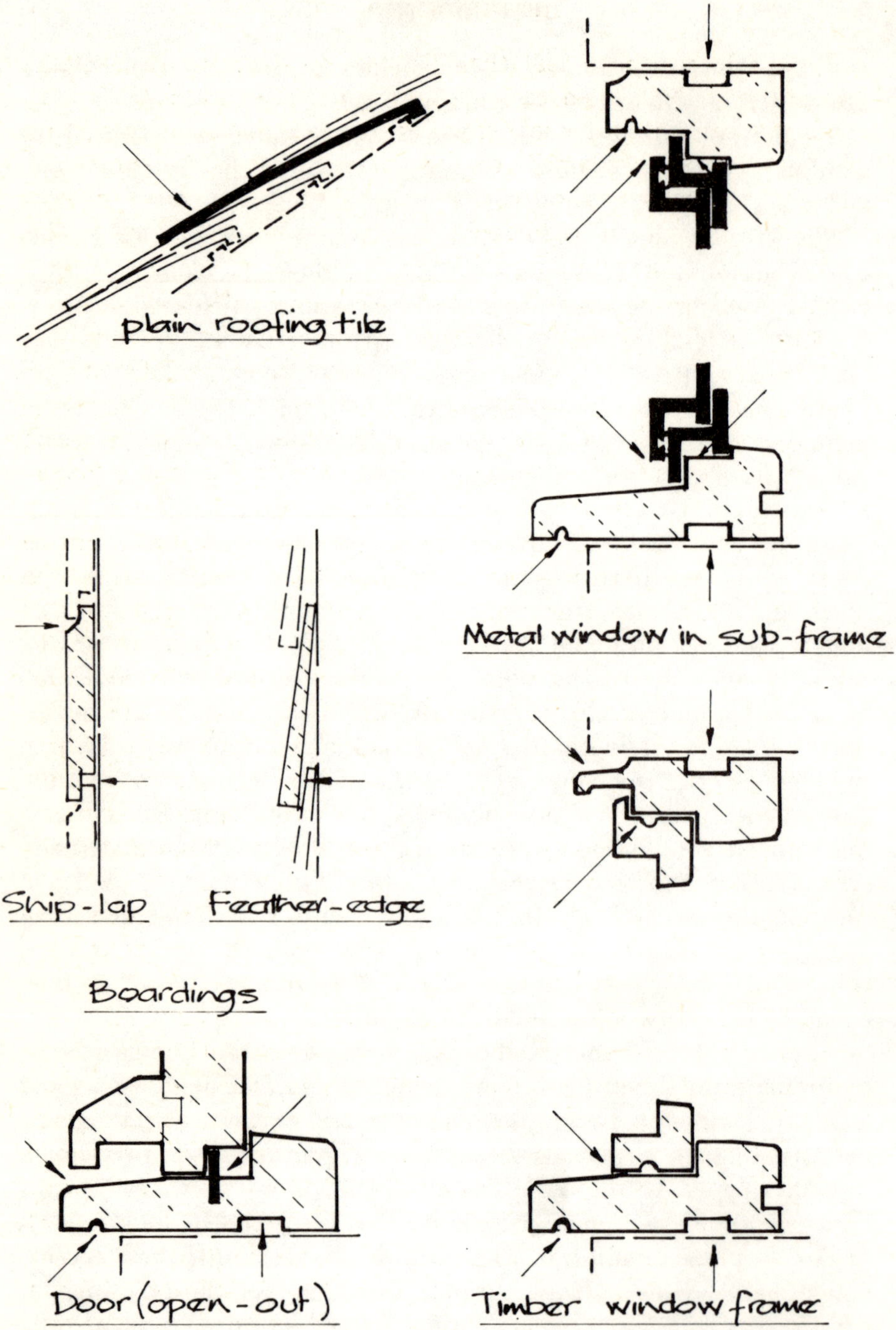

Fig 1.3 Anti-capillarity measures provided in manufactured components (arrows show breaks in capillary paths)

cavity should never be less than 25 mm in order to allow slight convection of the air. In cavity-filled insulation, this cannot occur and as a result the outer leaf takes longer to dry than normal. The type of fill used is 'foamed' to prevent capillary movement to the inner leaf. A better method is to increase the space between the two skins and to 'build in' rigid foamed sheet insulation bonded to the inner skin while still retaining a 25 mm cavity.

Rendering to existing solid brick walls or to lightweight concrete blockwork is an alternative in that it forms a barrier. However, it must be applied in two coats, with the finishing coat being thinner to reduce shrinkage of the cement during its setting action. It is also wise to stipple the surface since this tends to discard water more easily than a smooth surface, which can assist capillary movement.

Just as brickwork can attract water, so shrinkage of the mortar joints can provide paths into the surface. The joints should be struck to discard the water since, if it saturates the surface of the brick, it could lead to frost damage.

It is possible to treat the face with a clear silicone-based liquid spray, which dries to a waterproof film, but this is not advisable since it may crack with thermal or structural movement and any water that penetrates the cracks cannot escape by evaporation and will almost certainly cause spalling of the surface through frost expansion. Figure 1.4 shows details of the construction techniques discussed in preceding paragraphs.

Ground-water is the second obvious source of dampness and everyone must have heard of rising damp. This again is capillary movement, which can occur upwards through the brickwork, through fine cracks in the oversite concrete of solid floor construction, or at the joint between the wall and the oversite concrete. The combination of a damp-proof membrane to the oversite and a damp-proof course to the brickwork will prevent this if properly applied using the specified quality of materials. Adequate laps *must* be provided and they must be stuck together to avoid possible paths for the water to travel. Ideally the d.p.m. should be located below the oversite and dressed up the wall into the d.p.c. This allows for some settlement of the walls without putting tension on the d.p.m. A 'failed' d.p.c. should only occur because of severe structural movement and many so-called failures are really the result of bad workmanship where d.p.c.s have been butted instead of lapped at changes of direction or the start of a new roll. Damp-proof courses and the underside of thres-

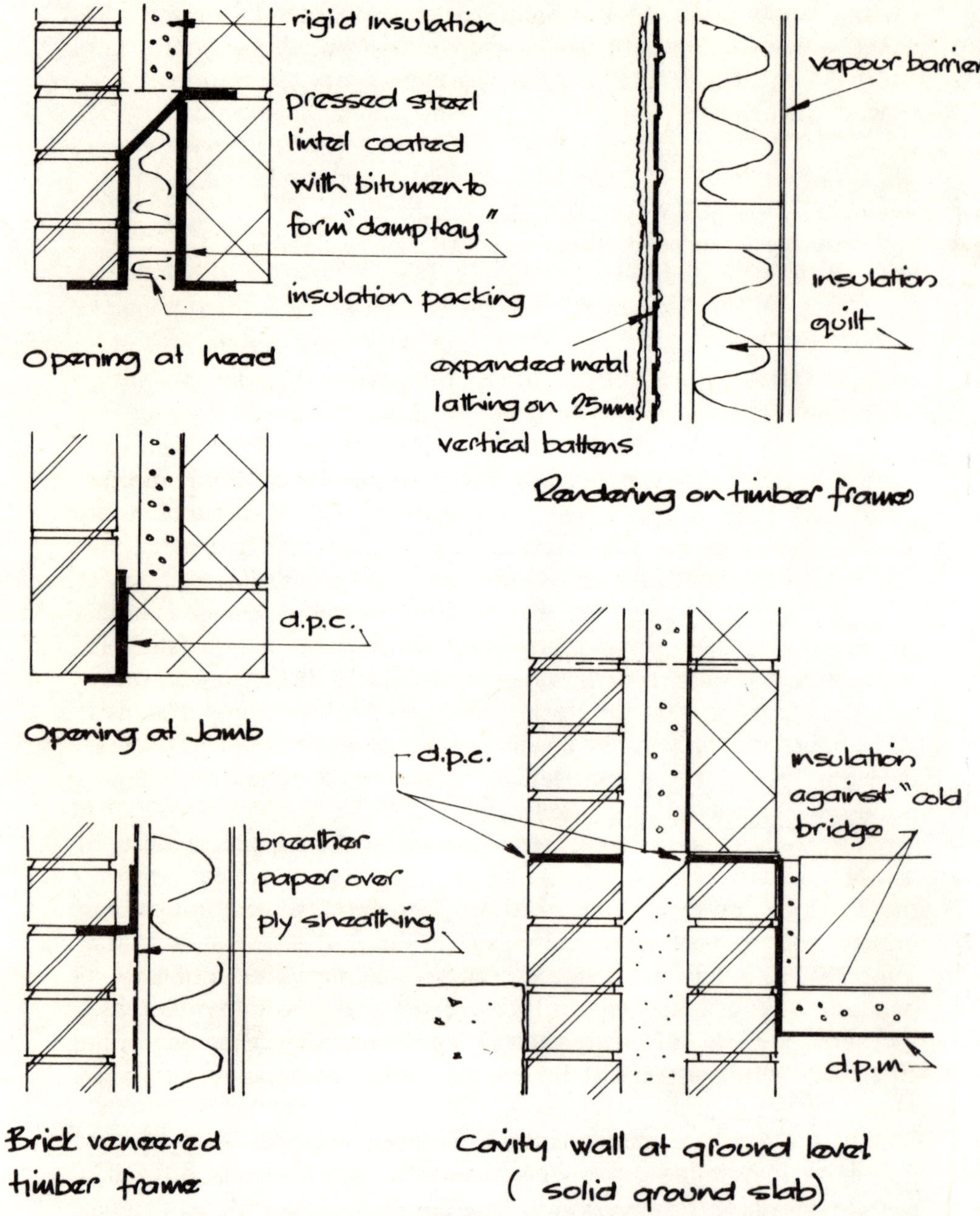

Fig 1.4 Constructional detail to prevent penetration of dampness into the building

holds to doors should also be located at least two courses of brickwork above ground level. There are two reasons for this:

(a) rain is unlikely to rebound more than 100 mm or so from a hard surface;
(b) surface water should not flood to more than about 75–100 mm in the event of storm drains being overloaded in a thunderstorm. This should normally ensure that brickwork above the d.p.c. is not saturated. In areas of known risk of flooding, such as those near rivers or open storm drains, it may be necessary to increase the height of the d.p.c. and to use a suspended ground floor.

Details of d.p.c. and d.p.m. installations can be found in *Building and Civil Engineering Construction* Volume 1.

The other causes of external water problems are less obvious and generally are caused rather than happen. Burst water mains are not an uncommon sight, though they should be, and are caused by the mains supply being too close to the surface of the ground. One contributory factor is traffic loading or traffic vibration, which can fracture the pipes, particularly if they are very old. A second is prolonged hot sunny weather, which can dry out clay soils and crack them quite severely, causing extensive movement close to ground level. This can overstress the pipework. Similarly, drainage pipes may be subjected to vibrations or thermal movement, but if properly bedded, using flexible joints, no damage should occur.

Many rain-water goods are plastic in present-day houses, and u.P.V.C. has a high rate of expansion and contraction. It is quite possible, therefore, to find joints in guttering or downpipes being displaced if not regularly checked. The maintenance programme should also include clearance of leaves, dust, seeds, spores and moss from the gutters and a check on the downpipes for blockages. If carried out each autumn and spring, there should be no problems. Cracked or overflowing gutters and downpipes can cause serious damage including wet rot in fascia boards and saturation in brickwork.

Other aspects of maintenance include inspection of roof tiles and the roofing felt (the latter from inside the loft) to detect, and where necessary make good, damage; inspection of external timber for signs of wet rot; and inspection of surrounds to the house to check that flower-beds or piled earth, sand, etc. have not breached the d.p.c.

Figure 1.5 shows diagrammatically some of the main areas of risk by damp penetration.

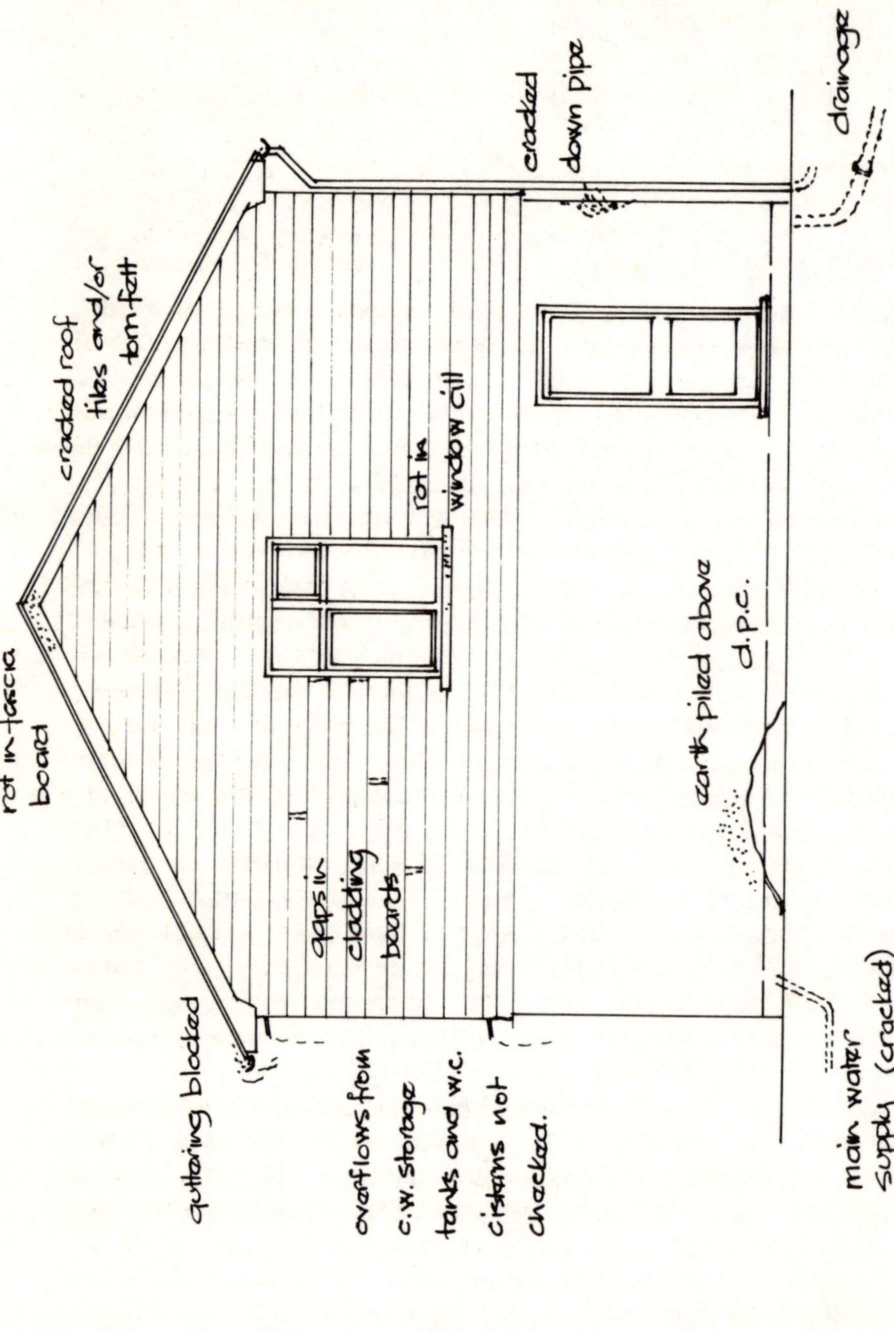

Fig 1.5 Areas of risk (external)

Internal sources, effects and protective measures
The least understood cause of dampness in buildings is condensation. People can accept that condensation occurs in kitchens and bathrooms because they can see the steam produced. When the water vapour cannot be seen, as in warm air, it is not so obvious that it exists. Chapter 6 deals more fully with moisture movement but, in simple terms, warm air holds more moisture than cold air. There are several contributors to the moisture in the air as well as that from boiling water. They include water vapour from gas hobs and flueless convectors, considerable water vapour from paraffin heaters and butane gas room heaters, and also the air breathed out by the occupants.

If unchecked, this air will pass into or through the fabric until it becomes cold, at which point the vapour condenses. With a solid wall this would be part-way through the wall, which is hot on the inside face and cold on the outside. Because the dampness dries out towards the warm face, many people think that it is caused by an external fault and it is often confused with rising damp at ground floor, or a leaky roof at first floor. It is neither of these: it is *interstitial* condensation (vapour condensing in the interstices, or inner pores, of the walling). If the fabric is not solid but is of framed construction (like a timber stud wall, a flat timber roof, or a conventional pitched roof), the condensation will occur when the warm air meets cold air or hits a cold surface. This dampness, particularly in the stud walling and flat roofing, will promote dry rot, which will eventually destroy the timber and affect other timbers within the building. In these conditions a vapour barrier is essential to prevent the passage of the moist air. Polythene sheeting may be used but it must be inspected to ensure that no tears are made during its installation or it will become useless. For any small amount that gets through, the ply sheathing is supposed to allow it passage through the breather paper and into the cavity. However, modern resin glues used in its manufacture may well make the plywood almost waterproof. The safest way to ensure a sound vapour barrier is to use, in addition, polystyrene-backed or aluminium-foil-backed plasterboard.

In pitched roofs the roof space should be ventilated since this will help to clear the water vapour before it has fully condensed. Ventilation will also help a little nearer the source of the water vapour; air changes are covered fully in chapter 6. There are certain situations where the water vapour is actually *attracted* to a surface. As an example, where there has been a case of rising damp the salts from the ground-water may have affected the internal plaster and made it *hygroscopic* so that it draws in water from the atmosphere. This could cause the occupier to think that the damp-proof repair has not been successful. To safeguard against this, all plaster within a metre of the known dampness from d.p.c. failure should be removed and only when the wall is fully dry should it be replastered.

Although thermal insulation has been the subject of increased interest over recent years, with wall, window and roof U values becoming ever smaller in new properties, it has served to emphasise locations where insulation is not very good, particularly when there is a sudden change from very good to very poor. These areas, because they present cold surfaces, attract quite severe condensation. They are known as *cold bridges* (where the cold from outside the building has bridged across to the inner surface) and occur typically at wall/eaves and wall/ground slab intersections. Some window opening lintels produce this effect and metal window frames can also be very cold, metal being a good conductor. The effect can be the development of mould growth or, in the case of windows, saturation of the window board, promoting wet rot. These effects can be reduced or even eliminated by detailed insulation of the fabric, and some manufacturers of aluminium window frames fill the spaces in the extruded frames with foamed polyurethane to reduce their thermal conductivity. Figures 1.6 to 1.8 show examples of bad and good detail.

A second, more readily visible, source of damp from within the building is in the leakage from internal supply and disposal services. This may be caused by electrolytic actions (covered in chapter 2), by thermal movement or simply by bad workmanship. Careful selection of compatible metals should reduce the risk of electrolysis and an allowance for expansion in lengths of pipework should overcome one aspect of thermal movement. Lagging of pipework likely to suffer very low temperatures, particularly in roof spaces above the insulation (but also in spaces under suspended floors), should counter the risk of freezing and splitting the water pipes.

As far as pipe assembly is concerned, compression fittings should be checked before assembly, to ensure that the ring washers are properly located, and then properly tightened; a visual check at a later stage will indicate leakages by the colour of the copper pipework (if used) close to the joint. Capillary fittings should be checked to ensure that the solder rings are present and a sufficient overlap at joints should be provided. Also, by using a reflective metal plate during fitting, this should ensure that the solder melts evenly around the jointing sleeve.

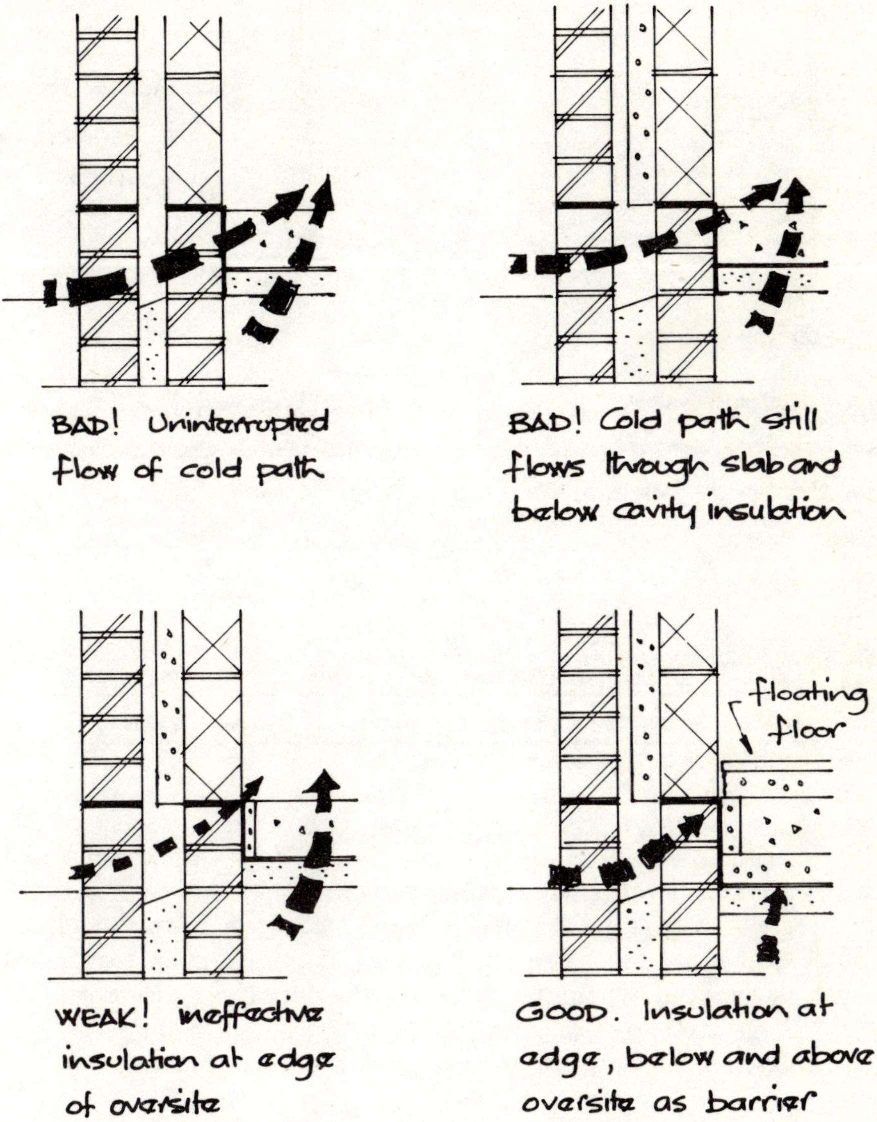

Fig 1.6 "Cold bridge" effects at junction of walls and solid ground floor slab

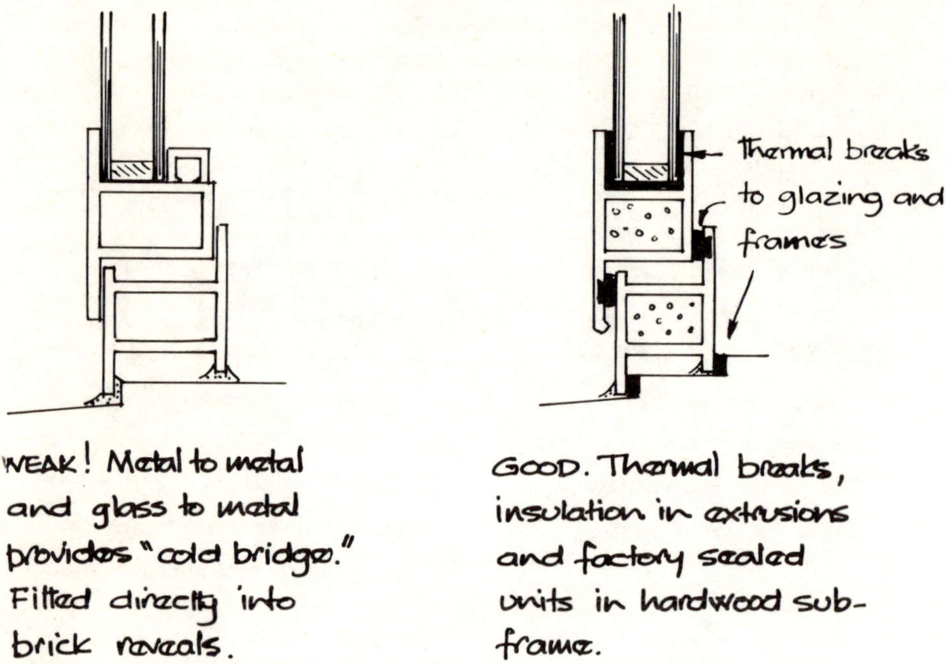

Fig 1.7 Typical aluminium framed windows with sealed glazing units (20mm gap)

Waste fittings, which are predominantly u.P.V.C., should be coated with sufficient jointing compound at connections (rather than fitted completely dry) and tightened carefully to avoid 'cross-threading', which could produce minute leakages. Supervision needs to be particularly keen when the joints to the supply or disposal pipes are hidden inside ducts or beneath floorboards, since they may not be detected sufficiently early to prevent the development of wet or dry rot.

Carelessness in use is the area least within the control of the builder since it will occur after completion. Typical examples are leaky fittings to supplies or wastes of washing machines and dishwashers, overflowing of twin-tub washing machines during 'rinse and spin' operations, wrongly located shower curtains and taps to sinks, or baths left running. The best the designer can do is to try to contain spillages in these locations. For example, kitchens and laundry rooms should have screeded and tiled floors of an impervious nature and the intersections with built-in cabinets should be sealed with a polysulphide sealant. This should also be used at splashbacks or sink units. Bathrooms and shower rooms on upper floors should either be screeded and tiled or, more realistically, covered with a seamless vinyl sheet floor covering, again paying detailed attention to joint intersections including shower tray to wall and bath to wall.

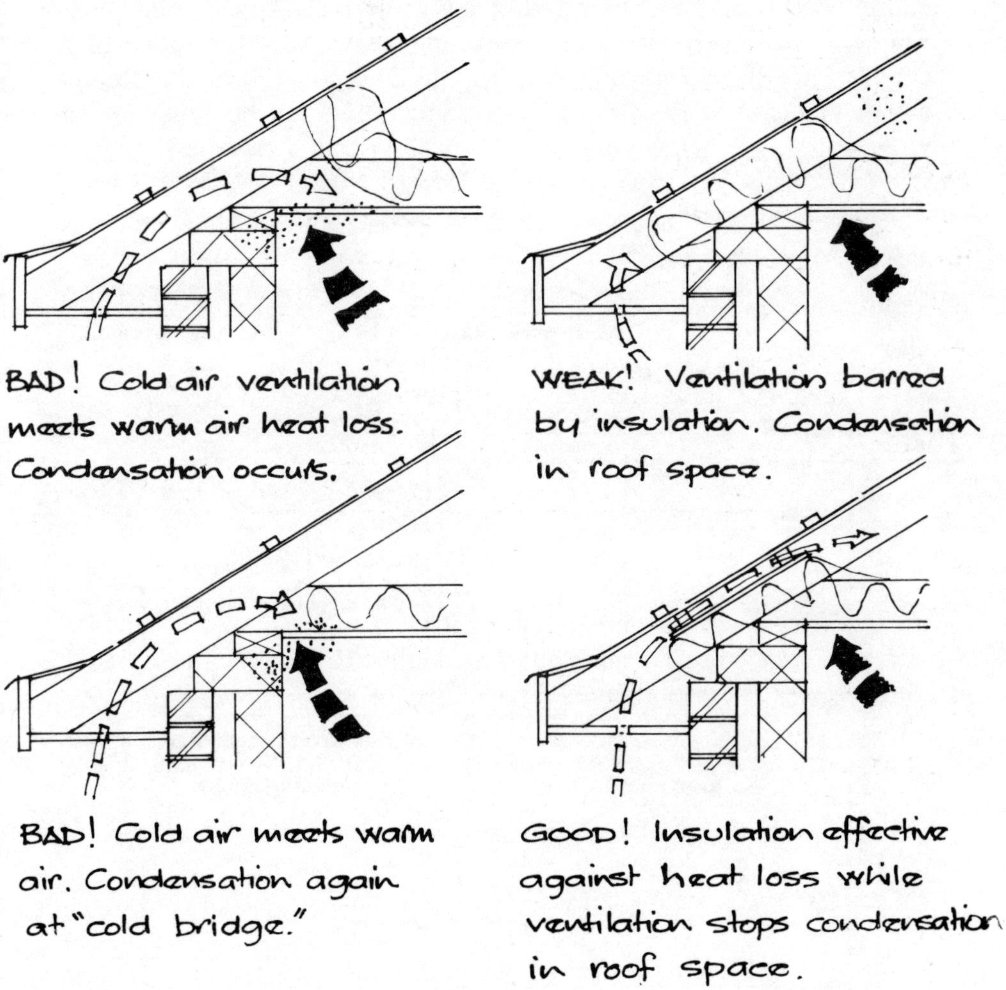

Fig 1.8 Problems created by inadequate or poorly located insulation to roof space

Walls should, of course, be tiled using an approved fixative and grout. Figure 1.9 shows typical problem spots inside the building with reference to damp penetration.

Structural deformation may occur, which results in fracture of d.p.c.s or the d.p.m. It might alternatively cause cracking in external renders or brick joints or possibly separation between two sections of a building. One cause is *differential settlement* (that is, where some parts of the building settle more than others). The most likely cause of this is poor foundation design based on inadequate preliminary site investigation. Thus ground heave or shrinkage, which should have been foreseen as possibilities, will fracture the foundations and distort the walling. A second cause may be alteration or extension work, which undermines the existing structure or is not

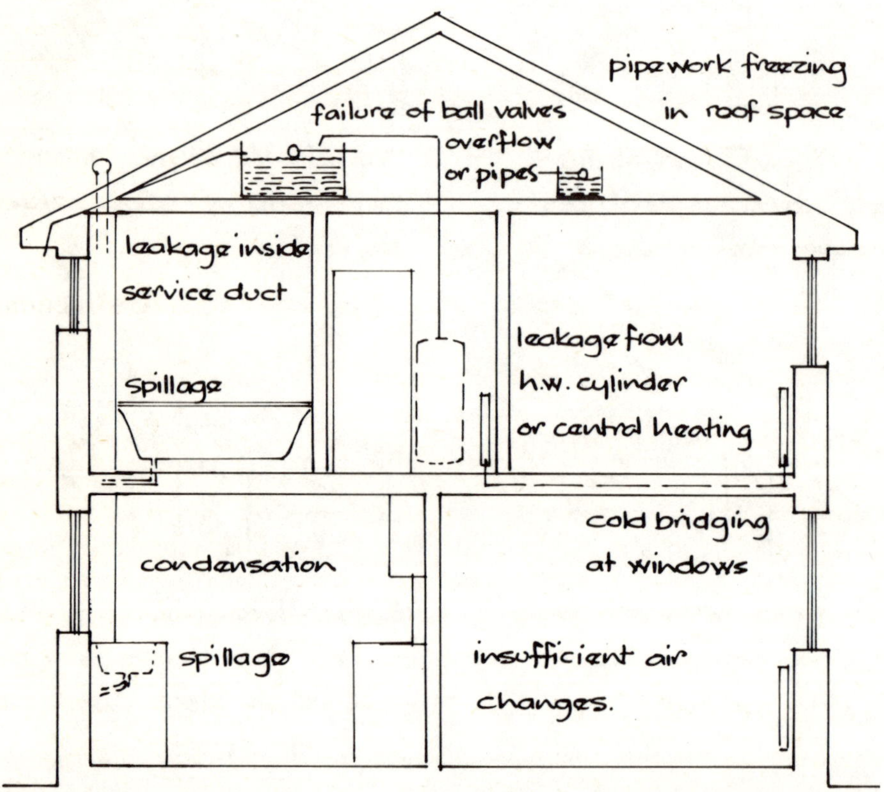

Fig 1.9 Problem spots causing dampness from inside the building inc. condensation

fully 'keyed in' with the existing structures. This can cause excessive deflection (in alteration work) or differential movement between parts of the building (in extension work), resulting in cracking of walls or separation between old and new work, allowing water penetration.

A third possible area of structural damage is in roof truss handling, manufacture (fairly rare), storage and assembly. Taking the least likely condition first: if a gang-nailed roof truss is manufactured with timber at too high a moisture content, the timber will shrink on drying out and the gang-nail plates will come loose. More probably, damage may occur during handling and/or storage, causing possible distortion of the members (or actual splitting) and, if stored in wet conditions, very high moisture content so that shrinkage movement can twist battens and dislodge tiles. Assembly might not be carried out to detailed instructions: if the trusses are spaced too far apart, the tiling battens will deflect, thus dislodging some tiles; if insufficient bearing at supports or lateral restraint is provided, there may be structural 'racking' or distortion. In all these examples the roofing fabric is disturbed, leading to rain penetration in its most vulnerable location.

Structural behaviour is dealt with more fully in Volume 2 and to some extent in Volumes 1 and 2 of *Building and Civil Engineering Construction,* but it is hoped that this chapter will have served to illustrate the need to develop a scientific approach to constructional problems.

Assignment 1.1

Consider a house known to you, or a case history, and establish where damp penetration from the outside has occurred. Ascertain the measures that have been taken to remedy the situation and whether they have been successful. From manufacturers' literature describe three methods of repairing d.p.c.s, including the scientific principles on which they are based, and determine the cost-effectiveness of each.

Assignment 1.2

Consider a house known to you, or a case history, where condensation has caused severe problems. List and describe the effects on the fabric and structure and explain how the problems have been overcome. From manufacturers' literature describe treatments to eradicate fungus on walls and protect them against future attack.

2 Materials Characteristics

Until relatively recent times the choice of materials suitable for building or civil engineering construction was confined to a small selection of those that occur naturally plus a few manufactured materials such as brick and iron. It is probably true to say that such materials are still favoured to a large extent today but the selection available, particularly in those manufactured, has expanded considerably.

In this chapter we shall consider the characteristics, such as workability, availability, performance in use, compatibility with other materials, cost effectiveness, weathering and useful life. There is always a temptation to investigate the microstructure and forms of bonding but we shall try in this chapter to concern ourselves more practically with the obvious properties since there are books that deal entirely with the chemistry and structure of materials.

We must always remember that we are examining materials in the context of construction and that what might be very important in other manufacturing technologies, such as the aircraft and automotive industries, is of little interest when related to a construction site. An understanding of materials behaviour and a sympathy with the 'working' of them on site are important technician functions, and although there are some aspects that are common to many materials, it is better to examine each for its individual characteristics.

CEMENT

Cement is probably the material most widely used in the construction industry and is readily available, not being restricted to any geographical region. It is a manufactured, rather than a natural, material, its main components being chalk or limestone, 80% (calcium carbonate); and clay or shale, 20% (silicate or aluminate). The dry materials are ground into powder form and mixed together prior to firing in a kiln. The firing process is progressive, first driving off the water (at around 400°C), then carbon dioxide from the chalk (at around 900°C), and finally causing a chemical reaction (at around 1350°C) between the calcium and the silicate/aluminate of the clay/shale. The resultant material is then ground to a fine powder, commonly known as 'dust', which will react when mixed with water.

The four compounds most common in Portland cements are shown in table 2.1, together with some performance characteristics. You will notice that one of the characteristics is heat output in setting, which is quite substantial. In chapter 4 we talk about heat energy and how it is stored when the state of a material is changed. In driving off the water from the mixture of chalk and clay we affect their crystalline form. This *water of crystallisation*, when reapplied to the finished product, begins to rebuild the crystals in a setting action, which releases the heat energy to drive off any excess moisture. The reactions continue over a period of time, up to about a year after their commencement, and comprise the setting action (initial and 'final') and the hardening of the material, this being the long-term reaction.

It is of critical importance to realise that once the setting action has *started*, the material must not be disturbed since any disruption by reworking the paste will render the material useless. It is therefore important to know the time limit between mixing and placing to avoid wastage and unnecessary expenditure. Because the reaction starts with the addition of water, it is known as *hydration* and, in producing heat, it is an *exothermic* process (an example of the use of scientific words to describe understandable actions, as described in chapter 1).

Previously we talked about 'excess' moisture, and it should be realised that the greater the excess the greater will be the shrinkage. Since this can cause cracking or unwanted moisture movement, it is useful to establish, by laboratory tests, an optimum *water/cement ratio* to suit the particular application.

The moisture movement is illustrated in fig. 2.1, which demonstrates the initial absorption creating a swelling action (increase in volume) followed by the initial drying shrinkage caused by the heat output. The C_3S and C_3A components form a gel at this stage and as the water is driven off, the gel becomes progressively more rigid. To slow down the effect of the shrinkage at the surface, the material may be 'wetted' at intervals. This process is known as curing and is

Table 2.1 Main compounds in Portland cements and their effects on the material

Name	Chemical formula	Short form	% content	Effect on setting	Effect on hardening	Heat (kJ/kg)	Moisture movement	Sulphate resistance
Tri-calcium silicate	Ca_3SiO_5	C_3S	45	Contributes to initial set	Contributes to early and late strength	504	Reduces with time	Not affected
Di-calcium silicate	Ca_2SiO_4	C_2S	28	Little effect	Contributes to late strength	252	Low	Not affected
Tri-calcium aluminate	$Ca_3Al_2O_6$	C_3A	10	Contributes to setting	Little contribution to strength	870	Continuous	Poor
Tri-calcium alumino-ferrite	$Ca_4Al_2Fe_2O_{10}$	C_4AF	10	Little effect	Low strength	420	Low	Not affected

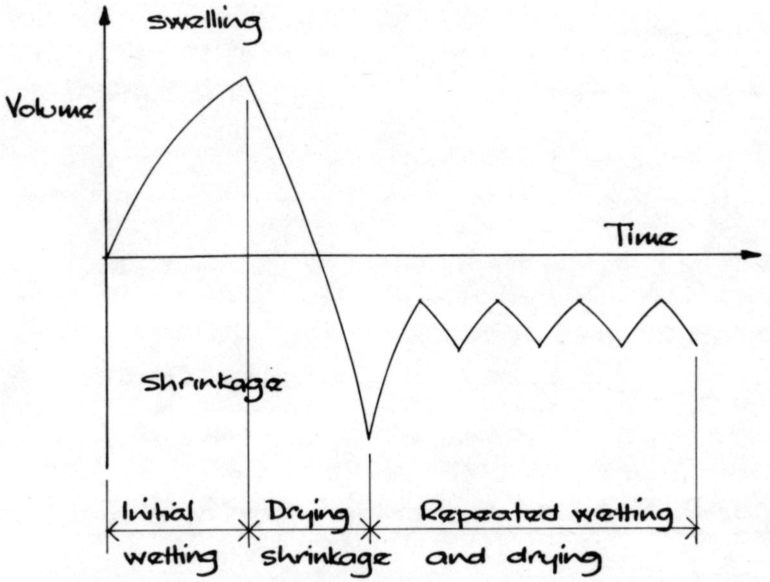

Fig 2.1 Moisture movements in cement

of particular application to cements used in rendering and concreting where large surface areas are exposed.

The development of material strength and the contribution of the main components during the hardening stage are shown in fig. 2.2 and it can be seen that the two major compounds, particularly the tri-calcium silicate, develop strength very quickly over the first 3 to 7 days after setting. They continue to develop but at a progressively slower rate and in theory the hardening process never stops.

Although we use the term 'Portland cement', it does not originate from that region but when set it has an appearance somewhat like Portland stone. The typical cement, described in table 2.1 and the above diagrams, is *ordinary Portland cement (O.P.C.)* and is used in a wide number of applications.

Other Portland cements, which have slightly different properties for specific applications, are as follows.

(a) *Rapid hardening Portland cement (R.H.P.C.)*, has a similar composition to O.P.C. but with an increased C_3S and a reduction in C_2S to accentuate the early strength characteristics. It is ground

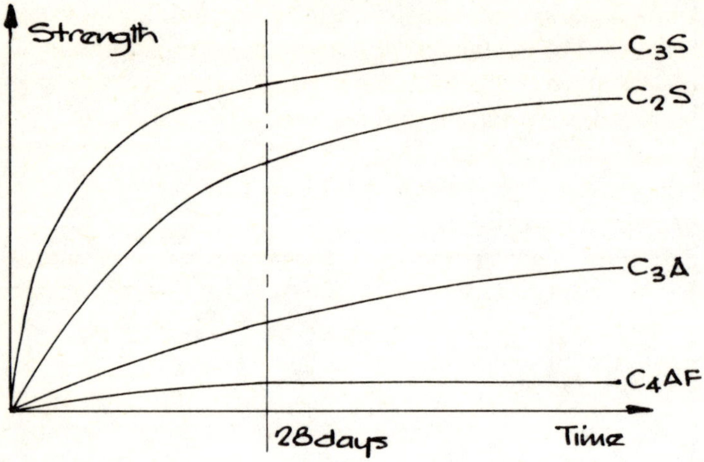

Fig 2.2 Strengths of compounds in cement

more finely and reacts more quickly to hydration but the increased rate of hardening should not be confused with its setting action, which is similar to O.P.C. Also, although its short-term strength is higher, its medium- to long-term strength is similar to that of O.P.C.

(b) *Low heat Portland cement (L.H.P.C.)*, which is also based on the same compounds as O.P.C., has an increase in C_2S and a reduction in C_3S and C_3A. Unlike the R.H.P.C., which gives off more heat in hydration, as table 2.1 indicates, the increase in C_2S means that the heat output is considerably reduced. This compound increase also slows down the development of early strength but again has no real effect on setting or medium- and long-term strengths.

(c) *Sulphate-resisting Portland cement (S.R.P.C.)*, has a reduced proportion of C_3A, this being achieved by an increase in iron oxide during manufacture, to reduce the attacks by sulphates in aggressive environments. A secondary effect is that the iron oxides make the colour, when set, much darker than that of O.P.C. These are not the only cements that are commercially available but they are probably the most widely used in routine applications.

Two cements that have particular applications but are rather more expensive than Portland cements are:

(d) *Pozzolanic cement*, which utilises volcanic ash rather than clay or shale. The crystals are much more rounded and mix more smoothly with the chalk/limestone component. The water/cement ratio required for workability can, therefore, be reduced.

(e) *High alumina cement (H.A.C.)*, which uses bauxite instead of clay/shale. The resultant effect is that the main compound produced is mono-calcium-aluminate, *not* C_3A, and there is very little silica in the gel. Its early strength is very high, much higher than R.H.P.C., but it has many disadvantages, which are well chronicled over the years, particularly since its abuse in the 1960s.

Cements are very rarely used in their pure form but are mixed with other cheaper materials to produce a variety of compounds, some of which are listed below. Many of these are covered in more detail later in this chapter and in chapter 4, but the list gives an idea of typical applications.

Concrete: for work of a structural nature. The cement is mixed with sand and gravel, which provide the mass. The cement provides the strength and its content varies from 10 to 25% of the mix, the most usual being about 15% by weight.

Mortar: for jointing brickwork, paving or precast components. The cement is mixed with soft sand and occasionally some lime, which again provide the mass. A 10% content is normal but 12.5% is not uncommon when stronger mortars are required.

Screeds: for finishing floor surfaces or levelling slabs prior to an additional finish. The cement is usually mixed with sharp sand, but for some finishes a crushed stone is preferred, and a normal cement proportion is around 10 to 11% of the mix.

Renders: for finishing, or making good, wall surfaces. Also they may be used for waterproofing of surfaces in basements, manholes or cess-pits. The cement is mixed with sharp sand as in screeds but provides a greater proportion of the mix, around 13 to 15%. A waterproofing agent is also sometimes added to the mix.

Plasters: for heavy-duty use, and sometimes cast components, where normal gypsum plasters are not sufficiently strong. Cement is added to the gypsum and in some cases strands of glass fibre are mixed with these components to produce glass-reinforced plaster (g.r.p.). The cement content is about 10% of the mix.

Fibrous cement: in flat sheet or corrugated sheet form for cladding or roofing of industrial buildings. The cement acts as a binding agent for mineral fibres, producing a fire-resistant material.

It can also be sprayed on to steel or pipework, where visual appearance is unimportant.

Concrete tiles: for roofing or wall cladding in a variety of profiles, textures and colours. Not really concrete, since only 'fine' aggregate is mixed with the cement, which contributes about 12% of the material. The result is a very durable weatherproof cladding component.

Wood-wool slabs: for roofing, or formwork to reinforced concrete. The cement acts as a binding agent for wood shavings to produce a medium-density strong boarding.

Grout or slurry: primarily a cement paste with some sand or similar additive. Used for packing around prestressing cables, surface jointing, making good of smooth concrete surfaces.

Blinding: an almost dry mix of cement and sand, about 8% cement, used to prepare irregular surfaces such as hardcore for damp-proof membranes. Also used to prepare ground for reinforcement to foundations or ground slabs, roads or pavements.

Reconstructed stone: for ornamental or cladding work. The cement performs a similar function to that when used in concrete but instead of aggregate a finely ground selection of stone and sand is used with 'white' cement.

This list, although not exhaustive, serves to confirm the statement that cement is the most widely used material in the construction industry.

AGGREGATES AND CONCRETE

Aggregates and concrete are discussed collectively since, although aggregates may be used on their own, they form the major component of all concretes. For normal *dense concrete* natural aggregates of sand and graded gravel are used in conjunction with cement. Crushed stone (limestone, sandstone or granite) may be used instead of gravel if a particular colour or texture of concrete is required, particularly if the surface is treated to expose the aggregate. Where no particular requirement is specified, it is likely that the aggregates used will be quarried locally to the site. This keeps the costs down, which is important because there is no sense in spending money that is supposedly for materials on high transportation costs.

Grading of aggregates used in concrete is important if a 'solid' material is to be produced, and although there is a nominal distinction between 'fine' and 'coarse' aggregates, the physical difference is that the largest particles of the fine aggregate will pass through a 5 mm sieve whereas the smallest of the coarse aggregates will not. The benefits of grading are illustrated in fig. 2.3, which identifies the problem of voids occurring in poorly graded mixes. The variety of aggregate sizes helps, if properly consolidated, to eliminate these voids and thereby to improve, or control, the strength of the concrete.

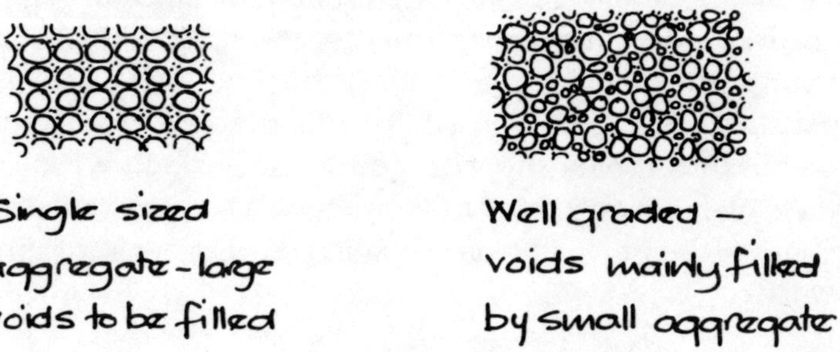

Fig 2.3 Aggregates in concrete

Ungraded aggregates are sometimes used deliberately when the formation of voids is required, but in these 'no-fines' concretes it is necessary to select the coarse aggregate for roundness of shape to improve workability and cohesion with the cement.

An important feature of any aggregate is that it must be clean and free from impurities to prevent the risk of an unwanted chemical reaction with the cement. When used in a dense concrete, the coarse aggregate must also have some inherent strength to contribute to the overall strength of the mix.

In order to reduce the need for rigorous analysis of aggregates on every mix, it is advisable to use a reputable source such as a well established quarry. Quite apart from the risk of chemical contamination, harmful to the cement, the aggregates must be free from dust, grease or clay, which might impair the bond with the cement. If the concrete is reinforced, salts in the sand could accelerate corrosion of the reinforcement and coastal sands should be washed thoroughly to remove all traces of sea-water.

Because both the fine and coarse aggregates may be washed, they will obviously retain some water which, if not assessed accurately, could upset the water/cement ratio of the mix. This will not only promote excessive shrinkage during the setting action and result in voids in the set material, but will also reduce the overall strength of the finished material (see fig. 2.4). A further problem of water content occurs in sand, which 'bulks' when damp. This means that it increases in volume in comparison with compacted dry sand or saturated wet sand. On a small site, the concrete or mortar may be batched by volume rather than weight and this could produce a concrete with insufficient sand content, causing small voids and reducing its strength, or a mortar with an imbalance of sand/cement, which might produce too strong a mix for the brickwork. An indication of bulking is shown in fig. 2.5.

Aggregates are not only used for manufacturing concrete. They may be used in combination with bitumens to produce road materials or on their own as sub-grades to ground works. They can also be used to strengthen poor load-bearing soils. Some typical applications are now listed.

Sands

Sub-grades for roads to provide level surfaces for concreting.

Bedding for paving slabs to pedestrian walkways or interlocking concrete blocks for semi-flexible roads.

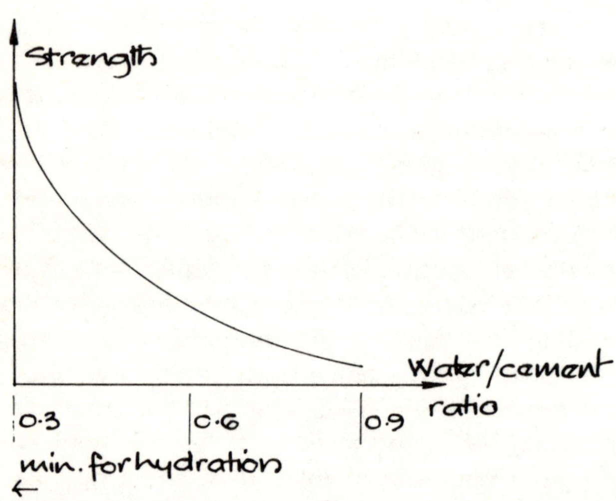

Fig 2.4 Variation in concrete strength with water/cement ratio

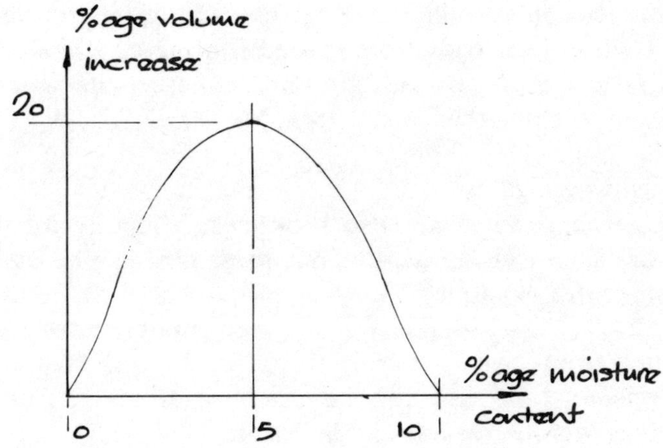

Fig 2.5 Bulking of aggregates

Dressing for cooling freshly spread asphalt on road or roof surfaces.

Dry jointing for dry-stone walls as boundaries or retaining walls in rural areas.

Also in conjunction with cement, as noted previously.

Fine gravel or shingle

Bedding and surrounding of drainage runs.

Surface-water drainage in open countryside or on embankments – misnamed 'sand drains'.

Dressing of road surfaces or driveways/parking areas.

Pebble-dashing: a form of wall finish to mask repair work. Also used as initial finish on some housing.

Solar reflection: white chippings on flat roofs to reduce the risk of roofing felt cracking or asphalt blistering or melting.

Tar macadam: mixed with pitch as a binder and hard rolled to provide a dense wearing surface for roads.

Also in conjunction with cement, as noted previously.

Large aggregates

Wearing courses and sub-grades for flexible road construction.

River and sea work: rock or boulder walls built up as jetties or quays.

Vibroreplacement: pillars of stone aggregate formed in soft soils to translate loads from spread footing to a firmer substrate.
Vibroflotation: layers of consolidated aggregate compacted by vibration to strengthen weak pockets of earth.

Lightweight aggregates

Lightweight aggregates are also produced, primarily for lightweight plaster or lightweight concrete, but these can also be used in other applications, for example:

Pulverised fuel ash (P.F.A.), an inert material, which is a by-product from power station furnaces, so fine that it is almost self-consolidating, and can be used for filling old underground workings or swallow-holes in limestone.
Bitag, a mixture of rounded lightweight aggregate particles and bitumen to form an insulating layer for flat roofs.
Vermiculite beads for thermal insulation of internal roof spaces, cavity walls and vertical service ducts.

Dense concrete

This is probably the most common form of concrete, being used for precast, prestressed and in situ concrete work (both plain and reinforced) and because of its familiarity it can be abused.

We have already discussed the principal components of concrete as being cement, fine aggregate, coarse aggregate and water. By careful selection of the aggregates and close control of the water/cement ratio, it should be possible to produce a strong, dense mix that does not shrink too much and develops strength after placing. A lot depends on the 'batching' of the mix and the actual mixing process. By varying the ratio of the three materials (not at this stage considering the water), we can vary the strength of the finished material. For most situations we can use a 'nominal' mix with a well-established set of limits within which to work. For special situations a contractor may, however, wish to use a 'design' mix, either to improve the strength or to reduce the cement content and thereby effect a saving in cost. Both types are dealt with in detail in the next chapter with regard to testing and quality control.

Concrete mixers come in a variety of output sizes so that even the smallest job can be mixed mechanically, usually by a gravity tilting drum mixer. Factory-produced concrete may alternatively be produced using pan mixers, and ready-mix lorries mix the concrete continuously while delivering it to the site. Although carefully selected aggregate, or pozzolanic cement, can help, the main contributor to the workability of the mix is the water content. This must also be carefully controlled and water should *not* be added to a mix just before placing to improve its workability.

When discussing cement we talked about 'initial' and 'final' set and, since the cement is the active ingredient in the mix, this is equally valid for concrete. In general, the initial set should not occur until about 45 min after mixing stops and final set should occur within 10 h of this time. What this means in practice is that the concrete must be transported, placed and compacted prior to the initial set taking place. Compaction may be achieved by one of three techniques:

(a) *manual*: either by tamping boards (for thin road slabs) or tamping rods (for test cubes);

(b) *vibration of the shuttering*: for factory-produced components or structural sections with a high reinforcement content. For this technique steel shutters are to be preferred.

(c) *poker vibrators*: which are immersed in the wet concrete after placing, and moved from place to place to ensure even compaction.

The prime purposes of compaction are to rid the mix of any air pockets, bubbles, etc.; to move the fine aggregate around the coarse aggregate; and to drive excess moisture to the surface, thus producing a dense mix. It is also used to circulate the concrete fully around any reinforcing bars or 'blockings' of polystyrene.

It should be noted that although the interaction of the cement and water produces heat in the setting process, this will not be apparent during the transportation and placing. Great care must therefore be taken when concreting in winter conditions to prevent the water from freezing prior to crystallisation. If this were allowed to happen, the frozen water would not combine with the cement and ice pockets would be formed which would seriously weaken the concrete. Protective measures such as insulated shuttering, warm water in the mix and propane air heaters will help to reduce the risks and straw insulation above the placed concrete, in the case of large exposed areas such as floor slabs, will further reduce the chances of freezing.

Conversely the surface must not be allowed to dry out too quickly since water being driven out from deep inside the cast concrete will need to escape. The surface must stay damp, therefore, to reduce shrinkage cracking. This is known as curing and is described in *Building and Civil Engineering Construction* Volumes 1 and 2.

Because there are so many variables, the strength of a mix, whether 'nominal' or 'design', will not be precise but *should* be consistent within limits. Rather than specify a minimum strength, rejecting any work that does not meet this, a statistical approach is used. This is based on 'cube test' results, whereby each batch of mixed concrete is sampled with cubes being made as the remainder of the concrete is placed. The cubes are stored under ideal conditions so that they can be tested at a later stage. When tested, the results are plotted to determine the distribution of cube strengths and the acceptability of the mix. This is dealt with more fully in chapter 3 but a typical distribution curve is shown in fig. 2.6, indicating the small acceptable percentage of cubes that may not reach the 'characteristic' strength of the concrete.

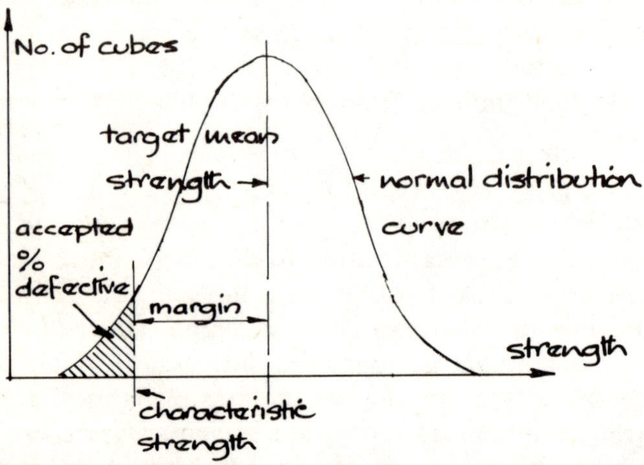

Fig 2.6 Distribution of concrete cube strengths

Dense concrete is used in a wide variety of applications and some of the more common ones are now listed.

Mass concrete: paths, ornamental blockwork, ground slabs, oversite concrete and foundations, gravity retaining walls, short bored piles.

Reinforced concrete: structural frames, columns, walls and slabs, beams, shell roofing, piled foundations, retaining walls, spread foundations, roads and bridges.

Precast reinforced concrete: cladding panels, portal frames, industrialised buildings, stair flights, caissons, permanent formwork, culverts and headings.
Prestressed concrete: bridge beams, large span slabs and beams, specialist portal frames.

Lightweight concrete
As the name implies, this concrete is 'lighter' or of a lower density than dense concrete. This reduction in density can be achieved in one of three ways:

(a) by eliminating the fine aggregate, which results in 'voids' as seen in fig. 2.3 and discussed previously. This 'no-fines' concrete has an irregular surface and does not require the same high-quality shuttering as normal dense concrete. It does, however, mean that the quality control in the mix needs to be of a high standard.
(b) by introducing lightweight aggregates such as perlite, pumice or clinker into the mix. The aggregate has trapped-air pockets, which improve the thermal insulation value of the resultant mix in comparison with that of dense concrete.
(c) by using selected fine aggregate and introducing air bubbles by 'entrainment', which are retained in the set concrete. This again improves the thermal insulation properties of the product.

The first two methods can be used in site-mixed concrete, which is placed in situ. The resultant concrete can be used structurally but cannot usefully be reinforced since protection of the reinforcement from corrosion is not as effective as with dense concrete.

The third method requires extremely close monitoring during the mixing and setting, which is best achieved in factory conditions. The components are therefore precast, normally in slab or block form.

Unlike dense concrete, which is water resistant when set and can, for some mixes, be totally waterproof, lightweight concrete is porous and needs to be protected from exposure to driving rain or ground-water. The lightweight blocks, using air-entrained concrete, have other advantages in addition to their thermal properties. They are easy to 'work' in that they can be cut or sawn cleanly, they are light to handle, and fixings are easy to install. Some typical applications are as follows.

No fines concrete: low-rise domestic buildings with very light imposed forces.

Lightweight aggregate concrete: low-rise domestic and commercial buildings of short- or medium-span floor bays.

Lightweight concrete blocks: domestic cavity wall construction (inner leaf) and partition or party walls. In this context the blocks are compatible with brick sizes, the dimensions on elevation being the equivalent of three brick courses in height and two bricks in length. The latest developments in this range of blocks are such that a cavity wall comprising an outer leaf of a 'half brick' skin, a 50 mm cavity and an inner leaf of 100 mm ultra-lightweight block provides a U value of just below 0.6 $W/m^2 °C$ yet retains a load-bearing capacity sufficient for normal domestic loading.

Additives

Additives are chemical agents that may be added to concretes, or for that matter screeds, renders and mortars, to alter the setting action or improve the performance. There are differing views as to their acceptability, particularly because of their uncertain long-term effects and the need to control their percentage content in the mix, and the safest course of action is *never* to use them in 'structural' concrete. They fall into five main categories, which reflect their applications, as follows:

(a) *Retarders*: these slow down the start of the initial set, which can be useful when laying very large plan areas of concrete for driveways, paths, oversite concrete and estate roads.

(b) *Accelerators*: which speed up the onset of the initial set and hardening processes. This can be useful in winter working conditions but should only be applied to mass concrete since the agent can promote corrosion in reinforcement.

(c) *Plasticisers*: which are particularly useful in mortars, screeds and renders since they allow a reduction in water/cement ratio with a consequent reduction in the risk of shrinkage cracking.

(d) *Waterproofers*: which are also useful in renders, mortars to brickwork in manholes and inspection chambers and in concrete cladding panels.

(e) *Colourants*: these are added to the cement to provide a range of colours in the finished concrete. They are particularly applicable to paving slabs, concrete bricks for fireplaces, garden walls and road pavings.

More information may be obtained from the Cement and Concrete Association (C. & C.A.).

BRICKS

In building, bricks have a long history of use, dating back to Roman times, and brickwork is still a very popular walling technique in present-day construction, providing texture, colour and character in a variety of treatments. Bricks fall into two main categories, clay and calcium silicate.

Clay bricks

Clay, when damp, retains water and can be moulded easily into any required shape. There are also many varieties of clay with regard to density, strength and colour. When pressed or moulded, the bricks are 'fired' in a kiln to extremely high temperatures. This drives out the water and, at the highest temperatures, vitrifies the clay (traces of sand are converted into a glassy material). In general the higher the firing temperature the stronger the brick.

Needless to say, clay bricks perform extremely well in fire conditions, being unaffected by both flames and high temperatures. Depending on the density, bricks are also water resistant, this being particularly true of 'engineering' bricks, which are extremely dense. Figure 2.7 shows diagrammatically the brick-making process in its various stages.

The texture of a brick, which is dictated by the clay used, will affect its weathering performance and the lighter density bricks can be prone to frost action. Because the surface is slightly porous, water may lodge in the brickwork. When the temperature drops

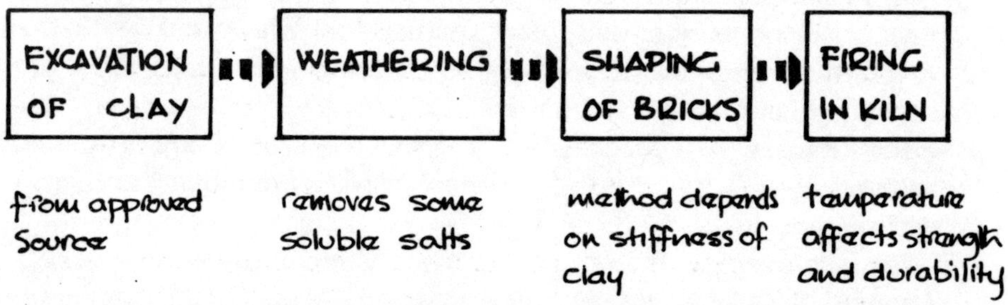

Fig 2.7 Stages of brick manufacture (clay)

below zero the water freezes, expands and fractures the surface of the brick causing a flaking (spalling) of the surface. Such bricks, therefore, should only be used internally. This problem can be aggravated if the mortar jointing is patchy, leaving pockets in which water can rest and from which it can soak into the bricks.

The shape of a brick will also affect the strength of the wall in which it is built. The lighter density bricks have a deep 'frog', to allow an adequate key for the mortar. This is not always fully filled, leaving weak pockets. Engineering bricks are shaped differently to overcome this fault. Either the frog is made much more shallow or it is eliminated completely by having vertical holes through the brick leaving top and bottom surfaces flat.

Clays often contain soluble salts, which are brought to the surface in the drying-out period when the water driven off in the setting action of the mortar travels along the surface of the bricks. This is called *efflorescence*, which is not harmful materially but may be visually offensive. The dried salts can, however, be brushed from the surface quite easily.

Some brick surfaces are improved by sand dressing, but these may be chipped off during the handling and laying of the bricks. It is more prudent to use 'facing' bricks with an inherent surface texture in external work, even though the initial cost may be a little more, since they are more durable.

Since the application of the bricks is almost totally for walling, the main consideration in the selection of the bricks is the location.

Commons or flettons: very basic and used mainly for internal work. Sand-faced flettons are sometimes used externally but are not a good substitute for facing bricks.

Hand made: this is something of a misnomer since it really means hand pressed rather than hand moulded. Because this process is expensive, only top quality clays are used, with the result that the final appearance is extremely pleasing. They are used for high-quality facing work.

Facing bricks: are made from selected clays and have a durable surface, less likely to be affected by frost action than sand-faced flettons.

Semi-engineering bricks: are of a higher density and higher strength than those already discussed and are suitable for medium-rise construction.

Engineering bricks: are of very high density and strength and so can be used for medium-rise construction but are more often used for work below ground, including basement construction as well as inspection chambers.

Calcium silicate bricks

Calcium silicate bricks are either sand-lime or flint-lime bricks, depending on the fine aggregate used with the lime. The aggregate and the lime are mixed using a very small percentage of water to effect the mix. The bricks are then formed under high compression using a hydraulic press and steamed at very high temperatures in an autoclave (steam pressure oven), which fuses the sand or flint at the surface. Because of this they are much more regular than clay bricks, their dimensions being to very close tolerances.

In comparison with clay bricks, calcium silicate bricks are much less susceptible to efflorescence but unfortunately can be attacked by salts below the surface in aggressive environments. This is known as crypto-florescence (yet another scientific word for a commonplace action) and can lead to spalling of the surface, exposure of the unfused core and eventual erosion of the brick. They should *not be used*, therefore, in coastal locations or heavy industrial areas. Frost resistance, however, is good and both types may be used externally. The bricks may be pigmented during manufacture to provide a range of colours. This is more usual with flint limes, which are generally preferred for facing work by virtue of their slightly textured surface. Sand limes are very smooth and are best suited for internal work. More information may be obtained from the Brick Development Association (B.D.A.).

TIMBER

As with brickwork, this is a well-established material with a long history of use both in building and civil engineering work. It has a huge variety of applications, including structural sections, temporary works, manufactured components and decorative claddings. Unlike the other materials discussed so far, it grows naturally and can be replenished by planting fresh woodland and forest (forestation).

There are two principal categories of timber, hardwood and softwood, but each category contains a wide variety of species with individual characteristics which make them suitable for specific applications. The obvious difference between a hardwood and a softwood tree is that the first is deciduous (sheds its leaves in the autumn) whereas the second is coniferous, and retains its leaves (sometimes called evergreen). A less obvious difference, which can only be seen under microscopic inspection, is the cell structure. Whereas softwood has only *two* main cells, one of which runs vertically to transmit sap, with the other running radially to store food, hardwood has *four* main cells. The hardwood has three sets of cells that run vertically, one of them much larger in cross-section than the others, and one that runs radially as with softwoods. The large cells, called vessels, are immediately noticeable under a microscope and on some hardwoods can be seen without magnification (see fig. 2.8).

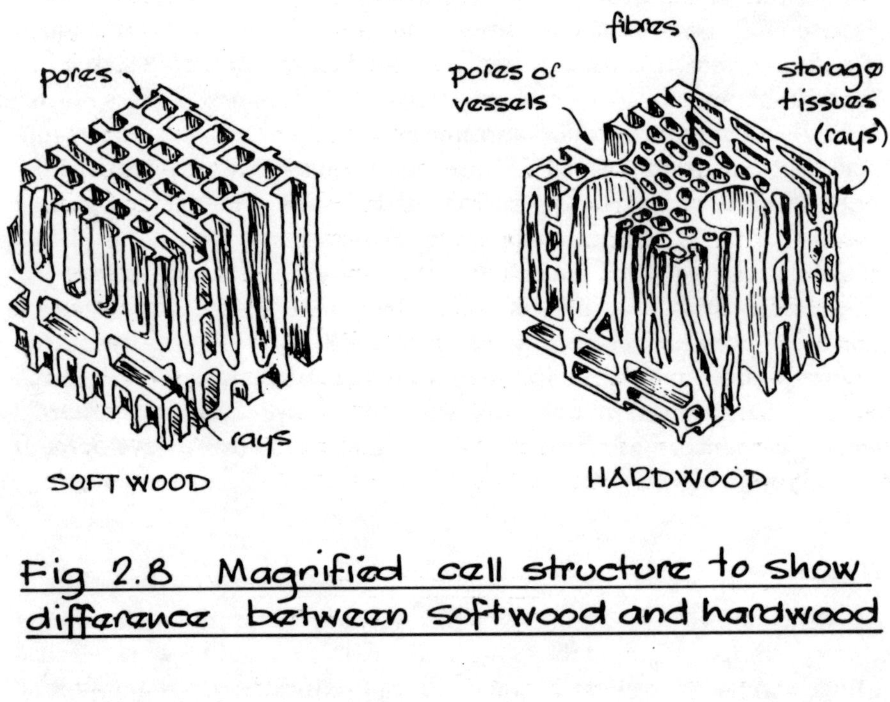

Fig 2.8 Magnified cell structure to show difference between softwood and hardwood

All the varieties of tree that we concern ourselves with in construction grow outwards, that is, the trunk of the tree builds up rings of growth each year. In the temperate climates of Sweden, Canada, Norway and the U.K., the growth pattern is quite marked and can be seen in the colouring of the annual ring. The cells in the spring wood are more open, to carry the sap, than those in the summer wood, which are more tightly packed. Because of its increased density, the summer wood is darker in colour. As the tree grows, the outer annual rings perform the primary function of carrying the sap, with the result that the inner rings compress and become less porous. These are called the heart wood (being at the centre) and the outer rings are known as the sap wood.

In the more tropical areas, where a huge variety of hardwoods grow, the growth pattern is not so marked since the seasons are not so obviously defined, and it is quite possible for two rings to grow per year. The heart wood is the most suitable timber for construction purposes, being less liable to moisture movement than the sap wood, but with timber prices rising all the time it is uneconomic to reject the sap wood. In fact any parts of the tree that can be salvaged are utilised in some way.

Having mentioned moisture movement it is as well to realise that, while growing, timber requires a large amount of moisture, and once it has been *converted* into a constructional form moisture can be problematic.

Moisture movement, the variation of moisture content causing the timber to swell or shrink, creates problems of dimensional stability in jointed components; for example, window frames stick through swelling or develop cracks through shrinkage.

Moisture content, the amount of moisture present in the timber, needs careful control both to reduce movement and to safeguard against rot. If the moisture is kept below 20%, the wood should not be liable to dry rot. Wet rot occurs at higher moisture contents, generally around 50%, but can develop at around 30% moisture content. The assessment and checking of moisture content is dealt with in chapter 3, as is the specification for location, but in simple terms:

$$\text{Moisture content (M.C.)} = \frac{\text{Loss in weight on oven drying}}{\text{Oven dry weight}} \times 100\%$$

Freshly cut (green) timber has an M.C. of about 130%, which makes it almost unworkable. In order to reduce this, the timber is dried out by seasoning and/or kiln drying. Cut timber stacked in an open mesh pattern, allowing air circulation, will lose quite a large amount of moisture by gradual evaporation. This technique has the advantage that without sudden movement the timber is unlikely to twist or warp. In kiln drying the process is speeded up and this is more likely to cause distortion. Because there is a constant demand for timber, some kiln drying has to be done but this should be the second stage after a large proportion of the excess moisture has evaporated naturally.

Rotting of timbers

When considering rotting of timber it is important to understand the difference between the two types. *Dry rot* occurs in unventilated conditions where the air is moist and temperatures are moderate, and where there is a ready supply of oxygen. It is caused by a fungus that feeds on the cellulose in the timber and when the cellulose has been eaten the cell walls collapse. The timber becomes dry and crumbles and can collapse quite suddenly. Because the fungus causing dry rot thrives in dark, unventilated locations it may not be seen. It can, however, be sensed by the smell it gives off and should be treated immediately to safeguard other timbers in the vicinity. Being a fungus it produces spores to spread its growth and these must be eradicated. They may even lodge in brickwork and this should be treated locally using a blowtorch after the rotted timber has been cut out and burnt to destroy the fungus.

Wet rot is also caused by a fungus but does not spread to the same extent as dry rot. It also requires a constant supply of moisture and is most likely to be found in timber close to leaky gutters, floorboards (and joists) in bathrooms where continuous spillage occurs, and on external timbers where a paint film has fractured and allows water to collect in the trapped area of timber. The effect of wet rot is also to break down the cell structure of timber but unlike the dry rot, which makes the timber very dry and whitish in colour, the wet rot darkens the timber almost to a blackish colour and the timber remains wet and very soft. This rot can be cut out, taking the timber back to its sound state without the same fear of spore spread.

The cause of both dry and wet rot is the fault of somebody. It is not truly accidental. It may result from the fact that the timber was not properly selected, the design/detail work was inadequate or the occupier of the property has shown extreme carelessness. If proper specification, design, constructional supervision and maintenance are carried out, rot should *not* occur.

Insect attack

The other risk of attack on timber is from insects. In the U.K. this is restricted almost entirely to *wood beetles*, and the four most common varieties are listed in table 2.2, which indicates the types

Table 2.2 Main varieties of wood beetle in the U.K.

Name	Approximate size of beetle (mm)	Approximate size of flight hole (mm)	Preferred types of timber and location
Powder post beetle	5	2	Sap wood of freshly cut hardwoods – timber yards
Furniture beetle (woodworm)	2.5–5	1.5	Softwoods and sap wood of hard-woods – house-timbers
House longhorn beetle	10–20	6–10	Softwoods – roof timbers, mainly in South-east England
Death-watch beetle	6–9	3	Sap wood and heart wood of hardwoods – old buildings, e.g. barns and churches

of timber and their locations. The adult beetles lay eggs in small cracks in the surface of the timber (possibly at joints) and when the grubs (larvae) hatch out, they burrow into the timber looking for food (starch contained within the cells). The grub stage generally lasts for about a year, which means that the grub will continue to burrow and feed at random until it reaches the pupa stage. This lasts a few weeks, after which the beetle emerges and chews its way to the surface to fly away in search of fresh wood in which to deposit its eggs. Unfortunately, it is only when the flight holes appear that visible evidence of infestation can be seen, by which time it is too late for that piece of timber. Most wood beetles are in the grub stage for about a year, but the death-watch beetle grub lives in the timber for up to 10 years and attacks the complete timber, eating into the heart wood and causing irreparable damage.

If this sounds rather frightening, readers should not be too alarmed. Each of the four beetles identified has its particular tastes and locations (as indicated in table 2.2) and it is in these conditions that particular precautions need to be taken. First, the beetles will not deposit eggs in preservative-treated timber, and secondly, the grub cannot survive in timbers that are fully treated. The difficulty lies with the death-watch beetle because hardwoods, such as oak, are not easily treated to any great depth. However, if the surface is regularly treated, it will deter the beetle from settling and laying eggs.

There are many types of preservative commercially available, in the following categories: (a) tar oil derivatives, (b) aqueous solutions of sodium or arsenates, and (c) spirit-soluble toxics. There is also a wide variety of application techniques, some of which are remedial. However, the most effective treatment is that carried out at the timber mill or yard so that the timber is protected at its earliest stage. It is also possible to treat large batches and achieve the maximum degree of penetration. Two of the more effective techniques are:

Vacuum pressure impregnation: The freshly cut lengths of timber are placed in a large pressure vessel and subjected to a vacuum pressure. This draws out moisture, which is removed. Pressure is then raised while preservative is pumped into the vessel, and then a second vacuum pressure is developed to draw out excess liquid preservative, leaving deposits on the cell walls.

Diffusion impregnation: The freshly cut lengths of timber are immersed in open tanks of the hot preservative solution. After a given time they are removed and wrapped in polythene sheeting to slow down the rate of evaporation. At a later stage the sheeting is removed.

With both these techniques the timber can then be seasoned in the normal way, as described earlier.

Applications

Application of use for timber is probably more varied than for any other construction material and is best considered by taking some of the more common species of timber of both types. Hardwoods, for instance, are more expensive than softwoods and tend to be used where appearance is of special importance. This is because they are often more attractive than softwoods and do not require as high a degree of decorative treatment. Softwoods, because they are cheaper, have much more widespread uses and, if treated with preservative, are competitive for durability. Typical applications are as follows.

Hardwoods

Structural: exposed framing, floor joists (oak), curtain walling (afrormosia, sapele, utile), river works (greenheart, teak).

Joinery: doors, shelving, fittings (teak, sapele, mahogany), stairs (ash, birch, teak, utile), windows (afrormosia, mahogany).

Temporary works: props, struts (oak, teak).

External claddings: t. & g. or shiplap boarding, fascias (sapele, mahogany, utile).

Internal panelling: mostly veneers on ply, chipboard or particle board (oak, teak, afrormosia, sapele, mahogany).

Second fixings: skirtings, architraves, panel trims (ramin, sapele, utile).

Flooring: mosaic block, t. & g. strip, parquet (afzelia, teak, afrormosia).

Softwoods

Structural: roof trusses, joists, timber framing (spruce, western hemlock, Douglas fir, Scots pine, redwood).

Joinery: door and window frames, shelving, cabinets (Douglas fir, Scots pine, pitch pine, larch, western cedar).

Temporary works: trenching, props, struts, shoring (spruce, whitewood, Douglas fir, western hemlock).

External claddings: roof and wall shingles (western cedar), t. & g. and shiplap boarding (western cedar, larch, Douglas fir, pitch pine, whitewood).

Internal panelling: boarding (Scots pine, western cedar, whitewood), ply, chipboard and particle board (whitewood).

Internal joinery: louvred doors (western cedar, redwood).

Second fixings: skirting, architraves (Douglas fir, whitewood, beech).

Flooring: floorboards (Douglas fir, western hemlock, Scots pine, whitewood).

This list is neither exhaustive nor exclusive and the individual tastes of the architect/surveyor/builder may determine choice. Also hardwoods and softwoods are often used in combination. This is particularly true of window and external door frames. Cills or thresholds of hardwood on softwood frames can be very effective.

The use of veneers on particle boarding is an illustration of the maximum use of waste timber. Offcuts of timber, or timber of poor structural quality, or badly warped sections, etc. are processed into small particles, which are bonded together and pressed into sheet form. The surface is sealed and prepared to accept a veneer of either whitewood or a selected hardwood, depending on its application. A plastic coating may even be applied for use in cabinet work.

More information on timber may be obtained from the Timber Research and Development Association (TRADA).

METALS

Metals have had a wide application in the construction industry for many years and although people tend to think primarily of structural steel in this context, the techniques of alloying, heat treatment, mechanical working and protection against corrosion are so well advanced that the range of applications is developing continuously. Metals fall into two main categories, ferrous (based on iron) and non-ferrous, and we can examine each in turn.

Ferrous metals

Iron is a fairly basic metal, being derived from impure mineral iron ore. The ore is 'reduced' by carbon (in the form of coke) and limestone (used as a flux) in a blast furnace at very high temperature. Impurities are removed by formation of a slag, comprising silicas (from the iron) and lime (from the limestone), which floats on the top of the molten iron. Further impurities are removed by blowing air through the molten iron. These include carbon and sulphur. The molten iron is then tapped off and cast into ingots of pig iron. This forms the basis of *cast iron*. Alternatively, the pig iron may be further refined by reducing its carbon content during a second melting process. More slag is drawn off as a result of the action of hot gases from the furnace passing over the surface of the molten iron. The iron produced is more workable and its carbon content is below 0.1%. This is *wrought iron*, so called because in its finished 'strip' form it is ductile and can be wrought (worked) into complex shapes for decorative effect.

A much more common product, however, is *steel*, which is produced by first resmelting the cast iron to remove all the impurities (such as silicon, manganese, phosphorus and sulphur) and then adding a controlled percentage of carbon. Other of the metallic elements may also be replaced but in a measured rather than a random fashion. The crystalline structure of the steel, and therefore its properties, vary with the carbon content and the temperatures at which it is worked. The three most common forms of carbon steel are:

(a) *Low carbon steel*: (carbon content up to 0.15%), which has fairly similar properties to those of wrought iron;
(b) *Mild steel*: (carbon content 0.15%–0.25%), which, although slightly less ductile, has an improved tensile strength and can be welded quite easily;
(c) *High carbon steel*: (carbon content 0.5%–1.5%), which is very hard but tends to be brittle.

Unlike cast iron, which has its carbon deposits scattered in small pockets throughout the metal, these steels have the carbon evenly dispersed, which makes them more suitable for forging, i.e. being made into hot-rolled, pressed or hammered sections.

The biggest problem with the iron and steels just described is corrosion by oxidation (rusting) and they require protective coatings to reduce the incidence of such attack. An alternative is to *alloy* the steel with chromium and/or nickel to produce *stainless steel*. This produces a corrosion-resistant steel, which does not need additional protection and has a wide variety of applications in construction work. For most stainless steels a content of about 14% chromium is used, but for special thin sheet work, such as sinks, a content of 18% chromium and 8% nickel is used. Alternatively, an alloy of copper can be produced to provide weather-resistant steel.

Applications

Typical applications for the irons and steels discussed are:

Pig iron: primarily as ballast for tower cranes, pile jacking and test loading because of its high density.
Cast iron: for mechanical components such as boiler casings.
Low-carbon steel: for some high-yield steels, rolled, extruded and forged sections, cables for ground anchors, suspension bridges and post-tensioned concrete, rivets and bolts.
Mild steel: steel plate, rolled sections, circular and rectangular hollow sections, reinforcing bars, pretensioned concrete. Used for structural frames, sheet piling, bridge construction.
High carbon steel: primarily for tool manufacture, cutting and drilling tools (occasionally with tungsten added).
Stainless steel: circular and rectangular hollow structural sections, hot and cold water pipes, sinks, trays and worktops for domestic

and industrial kitchens, reinforcement for thin concrete cladding panels, lifting and fixing bolts subject to severe exposure.

Non-ferrous metals

As the name implies, non-ferrous metals are not developed from iron ore but from other metallic elements, some directly and others by alloying a mixture of such elements. Their production is generally more expensive than that of irons and steels and, since those produced directly tend to be soft, they are not often used structurally. Most ores are smelted with carbon to reduce them to metals, the exception being aluminium, which requires electrolysis.

The most common base metals used in construction are aluminium, copper, lead, magnesium, tin and zinc, though not all of these are used directly.

Aluminium in its pure state is rather soft and is often strengthened by the addition of a small percentage of copper or magnesium. It is, however, often used in its pure state and is of much lower density than steel.

Copper is easily worked and can be used in sheet or tube form, having a wide variety of applications.

Lead is very soft and, although of high density, is easily worked. It does, however, have some limitations because of its toxic nature when in contact with water. For this reason it is no longer used in plumbing (the practice developed from working with lead – hence its name) in domestic hot and cold water systems.

Magnesium is used primarily as a toughening agent for aluminium, and *tin* similarly is used as an alloying agent with lead, either as the minor percentage in solder or the major percentage in pewter.

Zinc may be used in its pure form, in thin or thick sheets, or as an alloying compound. It may even be deposited, with oxygen, on to mild steel to form a protective coat.

The most common *alloys* with constructional application are brass, bronze, aluminium alloys and solder.

Brass is a mixture of copper and zinc and its strength and hardness depend on the zinc content. The lower the proportion of zinc, the softer the brass.

Bronze is a mixture of copper and tin and because a small amount of phosphorus is used during the alloying (0.1%–0.3%) is often described as 'phosphor-bronze'. As with brass, its strength and hardness depend on the secondary element to the copper. The greater the percentage of tin, the stronger the bronze.

The aluminium alloys and solder have already been described briefly and their applications, together with those of brass and bronze, will be discussed later.

Unlike ferrous metals, oxidation can actually form a protective coat from further weathering on non-ferrous metals. Aluminium turns grey, copper turns green, tin and zinc darken in colour. Similarly brass turns green and bronze turns greenish black. The problems of *corrosion* are from two sources:

(a) acids, alkalis or salts in rain-water or ground-water, or, in coastal locations, sea spray and mist.
(b) electrolytic action between metals linked by water (the electrochemical cell is described more fully in chapter 7).

With reference to the chemical action, this can be avoided either by using protective coatings or by not using particular metals in known aggressive environments, for example aluminium should never be used in coastal locations. The electrochemical action can be minimised by not linking metals whose positions in the galvanic series (table 2.3) are far apart. The table lists the metals in descending order of their rate of activity.

Table 2.3 The galvanic series

Magnesium (Mn)
Zinc (Zn)
Aluminium (Al)
Steel (Fe + C)
Brass (Cu + Zn)
Copper (Cu)
Bronze (Cu + Sn)
Stainless steel (Fe + C + Cr + Ni)

In simple terms an electrochemical cell comprises three components: anode, cathode and electrolyte. The more active metal (the anode) gives up its atoms into the electrolyte (the water), which may deposit them on the less active metal (the cathode). A common example of this in upgrading of existing properties used to occur where new copper piping to the hot and cold water supplies was used while retaining zinc/iron hot and cold water tanks. The zinc, being more active than the copper, corroded, leading to failure of the tanks and consequent leakages.

Not all electrolytic action is harmful and can be used to advantage if deliberately produced, as in anodising, electroplating and electrophoretic coatings, all of which serve to protect the metals from aggressive environments. Alternatively, if the risk of electrolytic action can be foreseen and the metals used cannot economically be changed, it is possible to introduce a third metal that is much more active. This can act as a sacrificial anode, which will need regular inspection to ensure replacement at suitable intervals.

Applications

The applications of non-ferrous metals, because they do not rust, are numerous and seem to increase as they are promoted by their manufacturers. Some of the more common constructional applications are as follows.

Aluminium: troughed sheeting for roof cladding, pressed 'sidings' for wall cladding, extruded sections for window and door frames (including secondary glazing), door and window furniture, in foil sheet form as thermal reflectors.

Copper: mini-bore and micro-bore pipework for central heating, hot and cold water pipes, hot water cylinders, plating or felt reinforcement (in expanded metal sheeting), in sheet form for roof cladding, printed circuitry for electrical control systems, wiring for electrical supply cables, decorative fitments.

Lead: in sheet form for roof cladding, for roof guttering, sound insulation, rainwater outlets; in strip form for 'leaded light' windows, sealing or jointing to prevent rain penetration; combination with felt as damp-proof course; to shield the effects of radiation (including X-rays).

Zinc: generally in sheet or wide strip form for roof cladding, flashings, roof gutters; in pressed or troughed sheet form for roofing; in deposited form for galvanising mild steel; in zinc oxide paints to coat mild steel components.

Brass: soft brass used for door and window furniture, ornamentation, light fittings; hard brass for plumbing connections, valves and machinery, bearings.

Bronze: high quality window and door frames, also toughened bronze for bearings; claddings, in thin sheet form.

Aluminium alloys: structural sections, curtain walling, pump components, lightweight scaffolding and stagings, ladders.

Solder: jointing work (particularly of electrical circuitry).

Metals are also used in electrical work such as transformers, batteries, etc. but this will be dealt with in chapter 7.

PLASTICS

The use of plastics, in comparison with the materials discussed so far, is still in the development stage, although such development has accelerated since the 1970s. These materials are predominantly the product of chemical research and it was only after their initial production that consideration was given to how they might be applied. This explains, to a large degree, the bad reputation that they had in early construction applications as being poor substitutes for well-established materials. This experience has, however, led to a refinement and more closely defined development of purpose-made plastics to fit job specifications.

Polymers are formed by joining together many small monomer units to produce very large molecules with high molecular weights. Natural polymers, such as cellulose (in timber) and rubber, do occur but the vast majority are manufactured using a carbon base. This has the advantage of being able to form chains and rings (see fig. 2.9). The processes of polymerisation depend on the nature of the monomers used, most of which are coal- or oil-based derivatives, and the main types of reaction involved are:

(a) addition polymerisation (producing thermoplastics);

(b) condensation polymerisation (by which thermosetting plastics are produced).

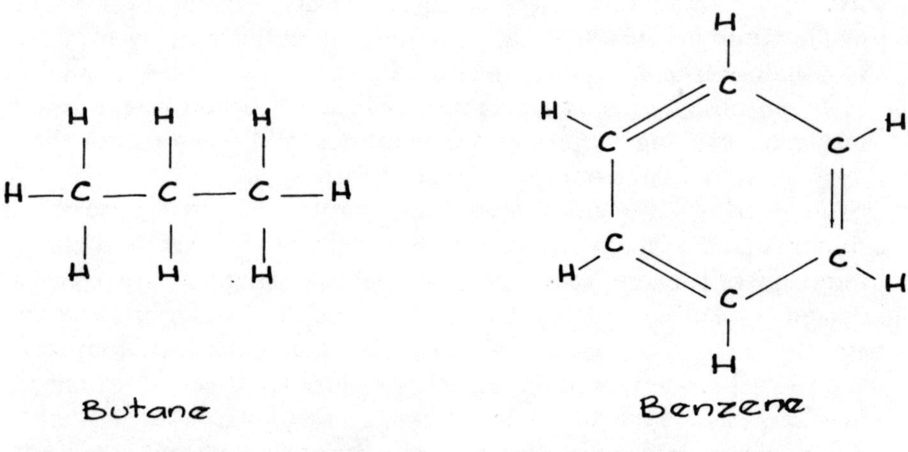

Fig 2.9 Chains and rings formed by carbon

Addition polymerisation is achieved using one material whose molecules are subjected to heat and pressure. The monomers link together to form chains. A typical example of this is polyethylene (commonly called polythene), where the monomers of ethylene join together ('poly' meaning many), to form the finished material.

Condensation polymerisation involves two completely different materials whose monomers are linked by water, which then condenses leaving the monomers forming chains which are cross-linked into a three-dimensional network. A typical example of this is melamine formaldehyde (commonly called Formica or other trade names), where the melamine molecules link with those of the formaldehyde to form a laminate.

Behaviour and applications

The density, texture and behaviour of plastics can be altered by addition, during manufacture, of solids, liquids or gases. Solids, in the form of 'fillers', may be used to produce a more economic material (a typical filler being sawdust) or to provide colour and/or texture. Liquids are often introduced to make the material more flexible and are known as plasticisers. Gases are often introduced to expand the materials to improve their thermal insulation properties. However, rather than concentrate on the manufacture, it is more important to consider the behaviour of plastics since this will generally determine their selection for constructional applications.

Thermoplastics soften when heated and, because of their normally soft texture, may be subject to severe surface grazing, which can provide traps for dirt to lodge. Some thermoplastics, however, are stable to higher temperatures than others.

Thermosetting plastics are more stable when heated, some being capable of resisting quite high temperatures, and being harder than thermoplastics are resistant to surface scratching.

The main enemy of plastics used outdoors is the ultraviolet component of sunlight, which can break down the chain structure, producing brittleness and loss of strength. Plastics are not usually affected by water, and are in fact often used as water and vapour barriers, but some 'plasticised' materials may be affected by the plasticisers becoming soluble and 'leaching out', causing brittleness. Their limited structural use is primarily because of their poor elasticity, though there are some notable exceptions which possess high elastomeric properties. In general, however, continuously applied loads will cause deformation. The prime cause for concern is behaviour in fire, since some plastics burn very easily and can contribute large quantities of heat, which accelerates combustion. Others give off toxic gases, which can be lethal. It is important, therefore, to check the full specification and test results under adverse conditions prior to selecting plastics for particular locations or applications. Some of the more common plastics and their applications are:

Thermoplastics

Acrylics: in resin form as adhesives, paint additives; in solid sheet form as decorative laminates and in baths, shower trays, etc.

Nylon: in solid form for door furniture.

P.M.M.A. (Perspex): in solid, coloured form in baths, shower trays, etc.; in translucent sheet form for lightweight roofing.

Polypropylene: water pipes (hot and cold) and fittings including valves and connections.

Polystyrene: best in expanded form as slabs of thermal insulating material.

P.T.F.E. (Teflon): sliding bearings (particularly bridges) because of low frictional properties.

Polythene: best in sheet form as damp-proof membrane (at ground level) or vapour barrier (walls and ceilings) but often used for protection and storage of materials; also for curing of concrete as protective sheeting.

P.V.A. (polyvinyl acetates): used in adhesives and paints but also sprayed in liquid form on to freshly placed concrete as degradable curing membrane.

P.V.C. (polyvinyl chloride): both in plasticised and unplasticised (u.P.V.C.) form; used extensively for rain-water goods, cladding, internal sanitation goods, external drainage pipes and inspection chambers; also provides sheathing for electricity cables, and additionally often used for floor tiles and sheeting, furniture upholstery.

Thermosetting plastics

Epoxy resins: as adhesives and binding agents in some floor screeds.

Melamine formaldehyde: heat resistant, high density worktop surfaces as thin laminate sheet bonded to particle boards.

Polyesters: primarily as fabrics for furnishing.

Phenol formaldehyde (Bakelite): mouldings for electrical fittings.

Urea formaldehyde: in expanded form, which is foamed into its location to provide thermal insulation.

Polyurethane: in foamed slab form often used in bedding and furniture as cushion material; in preformed expanded form as insulation to pipework; in solid form as synthetic rubber (see elastomers); in liquid form as hardening agent to paint films.

Elastomers: these are synthetic rubbers with high elastic properties. When cast into shape they will return to that shape after being subjected to compression. Used as weathering strips for cladding panels and window frames, (common trade name Neoprene). Also used in liquid form as basis of rubber paints.

This is by no means an exhaustive coverage of plastics and more information can be obtained from components manufacturers and the National Economic Development Office (N.E.D.O.).

PLASTERS

Plasters, in construction, are used almost exclusively as a means of providing a finish to walls and ceilings (as described in *Building and Civil Engineering Construction* Volume 1), both to make good any unevenness of brickwork, blockwork or concreting and also to provide a smooth, hard, hygienic surface that can be easily decorated.

There are some lime plasters still available commercially but the large majority of plastering is based on a material called gypsum. This is a mineral that occurs naturally and has the chemical formula $CaSO_4 . 2H_2O$. There are four main classes of gypsum plaster, the differences between them being dependent primarily on the amount of water of crystallisation they contain in their powdered form.

Class A is called a hemi-hydrate because in its manufacture much of the water of crystallisation is driven off. The heating process during the grinding of the powder reaches approximately 170°C, which dispels about three quarters of the water content, thus:

$$CaSO_4 . 2H_2O \xrightarrow[-1\frac{1}{2}H_2O]{170^\circ C} CaSO_4 . \tfrac{1}{2}H_2O$$

It sets by recombining with water, recovering its 'lost' water of crystallisation, to return to its original composition. This plaster, commonly known as plaster of Paris, sets so rapidly that it is not suitable for covering large areas. Its principal use is for mouldings and minor repair work.

Class B is also a hemi-hydrate ($\frac{1}{2}H_2O$), its manufacture initially following the same process as for class A plaster. However, to make the setting time more practicable, a small amount of retarder is added during the manufacturing process (an animal protein known as keratin). This slows down the recovery of the water of crystallisation, and therefore the setting time, making the plaster suitable for undercoat work and finishing plaster. When used as an undercoat, this plaster is sometimes mixed with sand. Alternatively, the plaster may be premixed (during its manufacture) with a lightweight aggregate. This has several advantages giving improved fire resistance and better thermal insulation. A more immediately noticeable advantage is ease of workability coupled with reduced drying-out time because of the reduced volume of active ingredient. For this reason lightweight plasters are extremely popular. The two main types are:

(a) *Bonding plaster* using vermiculite aggregate. This is used on *low* suction backgrounds, such as dense concrete, where the plaster 'solution' cannot be drawn easily into the surface pores.

(b) *Browning* plaster using perlite aggregate. This is used on *high* suction backgrounds, such as lightweight concrete blockwork, which have a porous surface.

One-coat class B plasters are now available, which may be used on most backgrounds and can be applied in fairly thick coats. These are popular since they reduce the labour content of plastering costs.

Class C is called anhydrous (no water) since during its manufacture the raw gypsum is heated to temperatures around 190–220°C removing almost all of the water of crystallisation. The setting process, as in all plasters, again involves the recovery of the lost water but this is a very slow process and so an accelerator is added during its manufacture (usually an aluminium salt) to produce a realistic setting time. This is still longer than that for class B plasters, which makes it suitable only for finishing coats. However, because less water is lost by evaporation, the finished surface is less porous and stronger than that of class B plasters.

Class D, known as Keene's plaster (and commonly known as Keene's cement) is similar to class C but is 'hard burnt' to about 400°C and ground very fine. The finished surface is almost glassy in appearance and non-porous. It is, however, little used nowadays because of its slow setting time.

What must be remembered with *all* plasters is that the drying-out process must be allowed to complete before decoration takes place. Since this drying out might also include that of the brickwork/ blockwork and the mortar joints (which can contain small pockets of water), it is sensible to add a few weeks to the manufacturer's specified times. If decoration must be carried out, *do not* paint the surface with a vinyl-based emulsion since the plastic film will prevent the excess moisture from evaporating evenly, resulting in small areas of bubbling and explosion through the surface.

Gypsum is also used in the manufacture of *plasterboard*, which comprises an aerated gypsum core encased in, and bonded to, specially prepared paper liners. This material is often used in preference to plaster, as 'drylining', to eliminate the drying-out period and allow for early decoration.

Assuming that sufficient time has been allowed prior to decoration, it should be realised that the surface, whether plaster or plasterboard, is porous. It should therefore be sealed, either with decorator's size or a proprietary sealer to minimise suction. This can also effect a considerable saving in quantity where paint is the decorative material.

It is also important when considering the selection of plaster or background preparation to follow the manufacturer's instructions implicitly, since failure of the plaster surface can be both embarrassing and costly, such failure being likely to occur after the building has been commissioned. Such instructions, together with a considerable amount of additional information on plasters and plasterboard, can be obtained from British Gypsum.

PAINTS AND COATINGS

There are two main reasons for painting or coating a surface. One is to provide a decorative effect. The second reason is to protect and/or preserve the surface being treated. Different paints or coatings are suitable for different types of surface, though some paints have a very wide range of applications. In addition to considering the material or surface to be painted, it is also important to realise the effect of its location. One obvious difference of location is the treatment for internal and external surfaces. Whereas internal surfaces have fairly static temperatures, external surfaces are subject to a wider temperature range with consequent expansion and contraction of the materials. Also external surfaces are much more subject to infra-red and ultraviolet light, which can affect the pigments in the paints or coatings. Attack from rain-water and waterborne chemicals can occur and grit in the air can settle on surfaces and then scour them in driving rain. Many paints or coatings that are suitable for internal use are therefore not recommended for use externally.

Paints

Let us look at the basic composition of paints. They consist essentially of the following.

(a) *The binder*: a liquid that dries to form a continuous film. This holds the paint together and should possess elastic properties to allow for movement of the surface on which it is deposited.
(b) *The pigment*: a solid that gives colour, stability and in some cases protection, as with lead or zinc oxide.
(c) *Thinners*: volatile liquids that allow the paint to be more easily applied.

Although these are the primary constituents, other materials can be added to modify the paint's properties. These include:

(d) *Extenders*: liquids that make the paint cheaper to produce, modify the gloss and improve application.
(e) *Plasticisers*: additives that reduce the brittleness of the binders when dried and help prevent chipping and flaking of the surface.
(f) *Driers*: catalysts that speed up the chemical reactions involved in the drying process.
(g) *Fungicides*: additives to provide resistance to mould growth.

Other additives can make the paint thixotropic, so that it forms a gel, producing a non-drip paint. A popular additive used to 'tint' paints is a stainer, which is a very finely graded pigment that disperses easily into the base pigment.

The process of painting involves two or more types of paint applied in sequence, i.e.

(a) *Primer*: this is applied to the unpainted surface to seal it and protect it from penetration, which might damage it (rot in timber or corrosion of metals). It also provides a key for later coats.
(b) *Undercoat*: this gives a visual covering power (opacity) and will fill any small holes in the surface, providing a key for

further paint. [*Note*: Some paints combine the properties of a primer and undercoat.]
(c) *Finishing coat*: this is applied for durability, finishing texture and final colour. It might be of gloss, eggshell, 'silk' or matt surface finish.
(d) *Lacquer*: this is a clear film that is sometimes applied to metal painted surfaces to provide additional protection to the finishing coat.

The actual techniques of applying paint vary, depending on the components or surfaces to be treated. Some examples are now listed.

(a) *Paint curtain*: a fine continuous curtain of paint is fed over a weir, covering panels passed through it at a controlled speed, the excess being fed back to the top of the weir. The panels continue past the curtain in a controlled environment where they are dried at slightly higher temperatures, which accelerate evaporation of the binders. Kitchen cabinets are often painted in this way.
(b) *Hot dipping*: a bath of paint is heated and the components are immersed so that they are completely covered. Metal window frames are sometimes primed using this method.
(c) *Electrophoretic dipping*: a bath of paint is charged with a small electric current and the component acts as a cathode on which the paint is evenly deposited. Steel cladding panels and some aluminium window frames are painted in this way.
(d) *Spraying and baking*: the paint is sprayed finely on to the surface of the component, which is then baked in an industrial paint oven to accelerate the drying and hardening process. This again is used primarily for metal components, which might include cladding panels, sheeting materials and pressed steel radiators.
With all these processes the environmental control is such that a dust-free, fully ventilated condition is assured. The bulk of painting in the construction industry, however, is done in situ, using brushes, pads or rollers, and the main types of paint used in these conditions are:

(a) *Oil-based paints* containing natural oils such as soya or linseed and/or synthetic alkyd resins. These paints dry in three stages: evaporation of the volatile thinners; chemical reaction by the oil and/or resin with the oxygen in the air; and finally polymerisation to give a cross-linked hard film.

(b) *Emulsion paints* containing resin (P.V.A.) or acrylic droplets dispersed or emulsified in water. These dry in two stages: evaporation of the water; and coalescence of the resin or acrylic to form the film.

These processes are shown in fig. 2.10.

Failure of paint films can be attributed to four major causes:

(a) *Incorrect selection of paint*: oil-based paints, if applied to bitumen-coated surfaces, cause 'bleeding' of the bitumen through the paint surface. Alternatively, the alkaline nature of freshly plastered walls causes saponification (soap forming) of the oil paints.

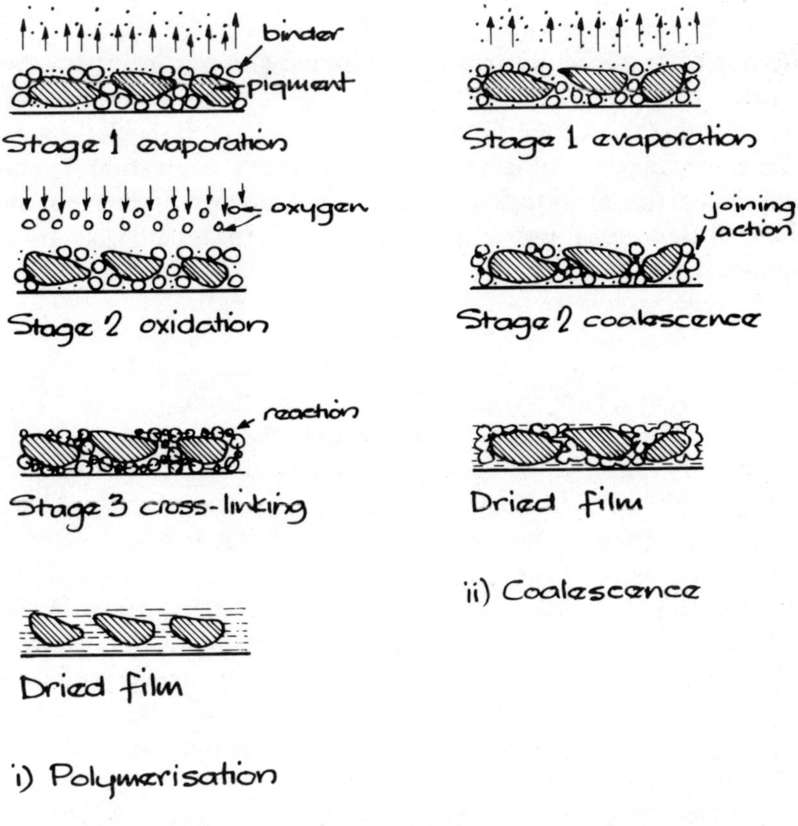

Fig 2.10 Processes of paint drying out

(b) *Inadequate surface preparation*: surfaces should be clean, non-greasy and free from other materials (such as dust) likely to affect adhesion of the paint coat. Alternatively, timber that has not dried properly can produce trapped moisture under the paint film, which expands with heat from the atmosphere or as a result of frost action, causing the film of the paint to break.
(c) *Bad workmanship*: the paint may be applied too thickly or too thinly. The painting may be carried out in damp atmospheric conditions (too early or too late in the day) or the coats of paint may not have been allowed to dry fully, causing 'crazing' below the finished surface.
(d) *Deterioration of the paint film*: either through sunlight affecting the pigments or the surface being scored and allowing dirt and moisture to penetrate the surface where it becomes trapped and blisters through expansion from heat or frost.

The last of these cannot really be avoided and repainting is generally required externally every 3 or 4 years. When this is carried out, the surface must be properly prepared by taking it back to a sound subsurface, since problems that are painted over soon re-emerge. This is particularly true of timber and an alternative coating to paint, which does not allow the timber to breathe, is to use a preservative stain.

Stains

Refined creosote preservative stains have been used increasingly since the 1960s and soak into the timber leaving an open-surface texture. A variety of colour pigmentations are commercially available, the darker pigments being more durable to bleaching by strong sunlight. In addition to shedding surface water, they also deter the natural oils in the timber from evaporating, and renewal is easily carried out by wire brushing the surface to remove grease and dirt prior to applying fresh coats of the preservative stain. Exposed end grains will need twice the number of coats as those applied to the long-grained surfaces and, in time, the wood will darken slightly with continuous reapplications. These stains will not, however, mask natural blemishes and the timber should be sanded mechanically to improve its surface texture. The consequences of neglecting periodic maintenance are, however, less serious than with paint films and can be remedied fairly easily.

Alternative types of paint

For other surfaces there are some alternative types of paint in common use, some of which are now listed.

Cement-based: sometimes known as 'stone' paints, these contain Portland cement and pigment in powder form, which can be mixed with water to form a slurry. Alternatively, the cement may be in suspension with vermiculite chippings in an oil or alkyd base, together with a pigment. Their main application is for external (and internal) renders. The water-mixed paints are, however, less durable in exposed conditions.

Rubber paints: based on chlorinated rubber and including pigments. Their prime application is in swimming pools but the white pigmented paints can be used to provide reflective paints to asphalted roof finishes.

Bituminous paints: which contain bitumen and solvents. Their prime use is to provide a waterproof coating for steelwork or concrete but they can be used to provide short-term remedial measures for cracked felted or asphalted flat roofing and to coat rain-water gutters and gullies.

As with plasters, it is important to follow the manufacturer's instructions and there is a vast range of technical trade literature available which covers much more than can adequately be stated in this chapter.

BITUMINOUS MATERIALS AND TARS

Bitumens

Bitumens are hydrocarbons that may occur naturally in limestone (when they are known as rock asphalt) or as deposits near lakes in Trinidad (lake asphalt). The majority, however, are obtained as a by-product in the refining of petroleum. They are used in road-building, in waterproofing of buildings and in coating of materials, as described previously.

All bituminous materials are thermoplastics and thus are affected by temperature change: they soften when heated and become brittle in conditions of extreme cold. The flow properties can be changed in three ways.

(a) *Cut-back bitumens* use a petroleum-based solvent, which reduces their viscosity. After application the solvent evaporates and the bitumen sets. The rate of curing depends on the volatility of the solvent. These are used in roadbuilding and impregnation of roofing felts.

(b) *Blown bitumens* are produced by blowing air through hot bitumen. Some oxidation and polymerisation occur to produce a harder, denser material suitable for flat roof construction.

(c) *Bitumen emulsions* are manufactured by using emulsifiers to disperse the bitumen in water (bitumen being insoluble in water). They set by evaporation of the water, followed by coagulation of the bitumen. They are used both in building and maintenance of roads, for waterproofing of basements and foundations, and for protection of timber and metals. Because they contain water they must be stored so that they are protected from frost action.

Tars

Other materials, often mistaken for bitumen, come under the heading of tars. They are derived from coal by distillation and have a higher specific gravity than bitumen but are often used for similar applications.

Road tar is coal tar adjusted by distillation to produce a dense, slow-setting liquid that can be used for surface layers of roads. It does, however, perform less effectively in this role than bitumen when subjected to hot summer temperatures and can become tacky. It is important not to overheat tars since they are more volatile than bitumen and should some of the vapours be driven off in the heating process the resultant tar may become too hard and brittle when fully set.

Low-temperature tar is thinner than road tar and is used primarily as a binding agent for stone chippings. The tar can be sprayed hot on to the road surface and covered with spar chippings, and the resultant mix is rolled lightly to form a surface dressing. It may also be used in this capacity to provide a reflective finish (using white or light-green spars) to flat felted roofing.

Other surfacing materials

An alternative coating for flexible roads is *macadam*, which incorporates graded aggregates. Macadam comprises graded 'coarse' aggregates, or, if an open texture is required, uniform sized coarse aggregates, which may be bound either with bitumen or tar. The macadam is laid in layers, each one being subjected to considerable compaction by heavy rollers. The resultant road surface is quite durable and considerably cheaper to produce than dense road tar. Although a tar-bound macadam may weather more quickly than one bound by bitumen, it has some advantages over the latter. It is less affected by oil spillage and performs better in heavy rain.

Mastic asphalt, as opposed to natural asphalt, is somewhat similar to macadam in that it incorporates aggregate. Because of its affinity to bitumen, graded limestone is often used as the 'filler'. However, red sandstone is sometimes used by preference. The bitumen, because it does not flow as freely as tar, is modified by using solvents to enable the compound to be mixed more easily. The resultant material is durable, waterproof and less liable to indentation than plain bitumen. It is, however, more expensive than macadam. It is particularly suitable for use on densely trafficked roads. If only fine aggregate is used, which makes it more expensive, mastic asphalt may be used for roofing and d.p.c.s.

Road asphalt may incorporate a small proportion of tar in addition to bitumen, to improve its fluidity, to reduce overall costs and to improve its performance in heavy rainfall.

Durability

The durability of bituminous and tar-based materials is affected by:

(a) *Extremes of temperature*: hot weather causes softening and if accompanied by warm winds (particularly on flat roofs) can cause the material to 'drift' so that some zones are thinned out. Cold winter weather can cause contraction and also hardens the material, making it brittle and liable to cracking. If water penetrates the surface and then becomes trapped when the material softens, closing the cracks, it can cause blistering. This can be catered for in part by using screed ventilators (see *Building and Civil Engineering Construction* Volume 1).

(b) *Application of stress*: being thermoplastics, bituminous and tar-based materials creep when loads are applied and can be squeezed out of joints. This is why polysulphides are preferred for expansion joints in roads and building surfaces.

(c) *Sunlight*: like many thermoplastics they crack when exposed to ultraviolet light. This is why reflective paint coatings or white spar chippings are used on flat roofs that use bituminised felt or asphalt. It is much better to provide tiling or paving to such surfaces since these provide a physical barrier to both sunlight and thermal movement.

(d) *Over-liberal use of solvents* which may not all evaporate. Those still active can cause localised softening and deterioration of the surface.

In the case of the manufactured materials, such as bituminised emulsions and bituminised roofing felts, it is critical that manufacturers' instructions are adhered to since failures can and do occur if short cuts are taken.

Assignment 2.1

Discuss the effect that modern constructional materials, particularly plastics and non-ferrous metals, have had on housebuilding, comparing a modern 'rationalised construction' house with one built at the turn of the twentieth century. In this context consider a modern house styled in the fashion of a 'Georgian' (or 'colonial') building, having upper elevations of shiplap boarding.

It is suggested that manufacturers trade literature, catalogues and specifications would be of help in answering this assignment.

Assignment 2.2

Consider the factory complex discussed in case study 8.2 of *Building and Civil Engineering Construction* Volume 1. List the metals used in the structure and fabric of the building, giving reasons why these metals may have been chosen. For each application describe the methods available for protection or decoration of the metal, stating what factors might affect the useful life of each component.

Assignment 2.3

Examine a road maintenance, repair or improvement programme being carried out locally, considering the choice of materials used. In the case of repair or maintenance, discuss factors that might have contributed to the deterioration and whether the materials chosen for the new work are likely to overcome those factors. In the case of improvement, discuss whether the materials selected for the new work are fully compatible with those already in existence.

3 Materials, Properties and Testing

In chapter 2 we considered materials with respect to their characteristics and constructional application. In order to ensure that such applications can be effective, it is necessary to ensure that their properties are consistent, a form of quality control, and that they meet specification requirements. In some cases this may be achieved by visual inspection but it is more likely that some form of testing will need to be carried out. It is in this area of quality control that the technician can be particularly effective, and he/she should be able to carry out such tests in accordance with relevant standards and interpret their results. The aim of this chapter will be to examine how the relevant properties of the materials discussed in chapter 2 can be measured.

CEMENT

Ordinary Portland and rapid-hardening Portland cements are governed by B.S. 12. The tests applicable are now described.

Chemical composition

Laboratory analysis shows the amounts of the elements present in the sample, from which the proportions of the four main compounds can be calculated. Setting rates, development of strength and resistance to sulphates are all affected by variations of composition.

Fineness

The speed of hydration reactions depends on the surface area of the cement particles: the smaller the particle, the more rapidly the cement will hydrate. The specific surface (total surface area per unit mass measured in square metres per kilogram) is found by measuring the flow of air through a plug of cement powder. In B.S. 12 are given minimum specific surfaces for O.P.C. (225 m^2/kg) and R.H.P.C. (325 m^2/kg).

Strength

The strength of cement will obviously affect that of concrete that incorporates it. The compressive strength of cement can be found by using either mortar or concrete cubes made from the cement. In either case, the cubes must use standard proportions of materials mixed in a prescribed way. The cubes are stored under controlled conditions of temperature and then crushed by forces applied at a standard rate. The standard gives not only the required minimum strengths but also the limits for the spread of results from a particular set of cube tests.

Setting

In practical terms cement mortars or concrete must not set too quickly after mixing, to allow for handling, transportation and placing. Cement must, however, develop strength at a reasonable rate, when used in concrete, to allow formwork to be removed fairly quickly. The setting test uses the Vicat apparatus shown in fig. 3.1. Neat cement pastes are made and when one of a standard consistency has been produced, the initial set is found using the $1mm^2$ needle. This is the time taken from mixing until the needle does not penetrate to within 5mm of the base of the mould.

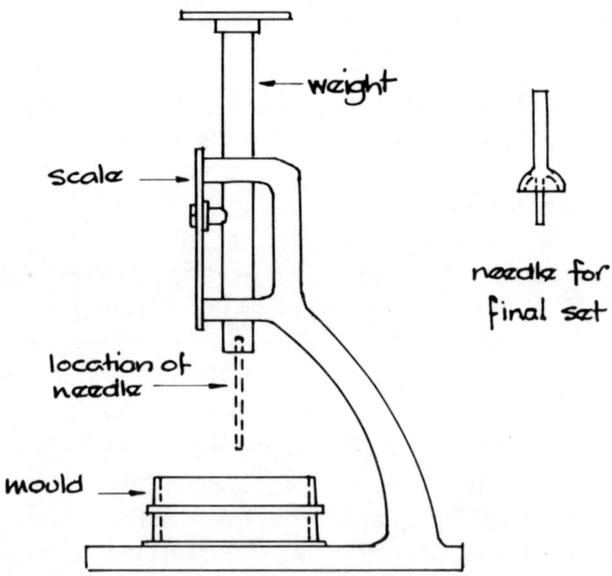

Fig 3.1 The "Vicat" apparatus

The final setting time is when the needle, but not its flange, makes an impression on the cement. A *minimum* initial set of 45min and a *maximum* final set of 10h is required by the standard.

Soundness
Occasionally during the manufacture of cement, some of the lime (CaO) does not combine with the other materials. This unburnt material could react with the water, producing swelling likely to cause loss of strength in the cement paste. Although the process would normally take a long time to show itself, an accelerated test can be carried out using boiling water. The Le Chatelier apparatus, shown in fig. 3.2 is filled with a cement paste of standard consistency as before. This is allowed to set, and the distance between the pointers is measured. The mould is then placed in water, which is brought to the boil, held at that temperature for an hour, and then allowed to cool. The distance between the pointers is measured again, and if the expansion exceeds 10mm, this indicates that the cement is unsound. However, because cement manufacture is subjected to a very high level of quality control before it reaches the customer, it is likely that it will leave the cement works in a satisfactory condition.

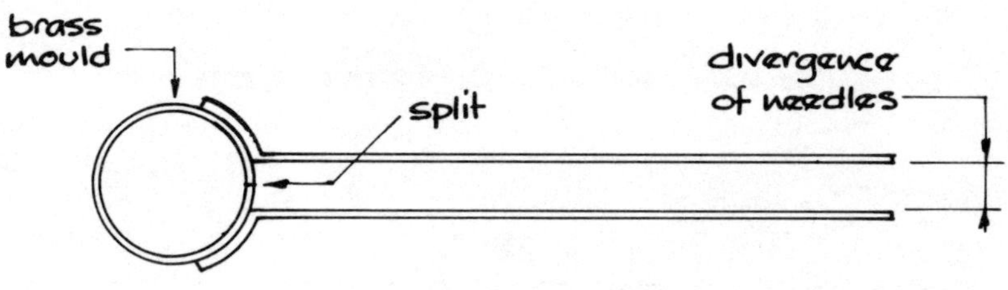

Fig 3.2 The "le Chatelier" apparatus

The performance of cement is more likely to be affected by poor storage, possibly at the builders' merchants, or, more likely, by the customer. The cement should always be stored in dry conditions where it cannot absorb water accidentally; stores should be rotated to ensure that cement used is always reasonably fresh; and the bags should not be stacked too high to avoid the risk of a compaction set. Also, bags should not be opened until immediately prior to use, since moisture absorbed from the atmosphere can lead to 'air hardening', giving a lumpy cement that is unlikely to develop maximum strength.

AGGREGATES

The importance of particle size distribution has been explained in chapter 2. Distribution can be assessed by the following methods.

Grading

Grading is carried out by passing a sample of the aggregate through a set of B.S. sieves and assessing the amount retained on each. The sample should be representative, being taken from different places in the stockpile, and either 'quartered' or passed through a riffle box. The cumulative percentage passing through each sieve is then calculated, the results being plotted on a graph to give a grading curve. There are four zones into which *fine* aggregates are classified. Zone 1 is the most coarse and is not usually employed in concrete mixes, which use large aggregate as well. Zones 2 and 3 are less coarse and are most suitable for concrete mixes (they are described as sharp sand), and zone 4 is the finest grade (soft sand), which is used for bricklaying mortars. Although fine aggregates for structural concretes should fall within the zones noted, it is quite possible to produce sound non-structural concrete using a non-standard sand. The results for a graded sample are shown in table 3.1, with the zone 2 limits shown alongside the sieve results. (*Note:* 1 μm = 0.001 mm.)

Table 3.1 Sieve analysis of a fine aggregate

B.S. sieve mesh size	Mass retained (g)	% retained on sieve	% passing sieve	Zone 2 limits
5 mm	15	5	95	90–100
2.36 mm	15	5	90	75–100
1.18 mm	30	10	80	55–90
600 μm	90	30	50	35–59
300 μm	90	30	20	8–30
150 μm	45	15	5	0–10
Base tray	15	5	*N*	*N*
Total	300	100	*N*	*N*

Coarse aggregates can also be graded using the upper range of B.S. sieves. These overlap with those listed in table 3.1 and are: 37.5 mm, 20 mm, 10 mm, 5 mm and 2.36 mm. As with fine aggregates, there are limits within the percentage passing each sieve but *none* should pass the 2.36 mm mesh since it could not be properly classified as *coarse* aggregate. For most structural concretes, a maximum aggregate size of 20 mm is used so that 95–100% should pass through the 20 mm mesh. In some cases, however, such as ribbed 'hollow pot' flooring, a smaller size of 10 mm may be specified.

Strength

Aggregates from known, approved sources are usually sufficiently strong for most purposes. Sometimes, however, crushing strength may be specified, as may resistance to impact or abrasion. These properties are primarily of interest in highway engineering and standard tests are available for measuring them.

Moisture content

As mentioned in chapter 2, the moisture contained in aggregates can affect the water/cement ratio of a concrete mix and there are several techniques that can be used to determine the moisture content of fine or coarse aggregate. The most simple method is to take a sample of the damp aggregate and record its mass. Place it in an open pan, as shown in fig. 3.3, and heat it gently until the surface is visibly dry. It is then cooled and reweighed so that the moisture content can be calculated thus:

$$\%\ \text{moisture} = \frac{(\text{Mass of wet sample} - \text{Mass of dry sample}) \times 100}{\text{Mass of dry sample}}$$

Having established this, the water content can be determined for the concrete mix proportions and taken into account as part of the water/cement ratio requirements for the mix. However, because this method may not be sufficiently accurate for fine aggregates, an alternative test, using the 'speedy moisture tester', can be used for both coarse and fine aggregates. This apparatus is shown in fig. 3.4 and makes use of the fact that calcium carbide reacts with water to produce acetylene gas. An aggregate sample of known mass is placed in the vessel along with a separate measure of calcium carbide and sealed. The two are then mixed and the pressure of the resultant gas is measured to give the water content.

Other methods, such as the 'siphon can' and 'Gammon–Morgan cylinder', may be used but these require an initial drying test on specimens of the aggregates prior to calibration. They are more usefully used for large quantities of stored aggregates, which may vary considerably in moisture content over their time stored on site.

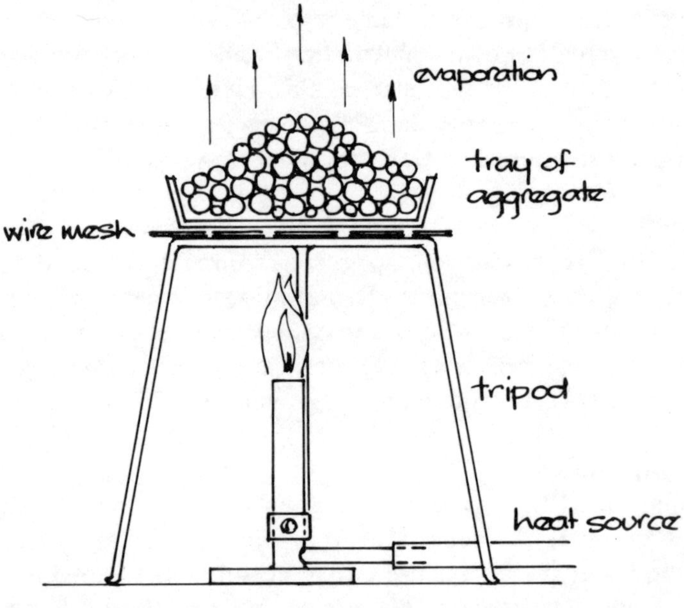

Fig 3.3 "Frying pan" method for determining moisture content of aggregates

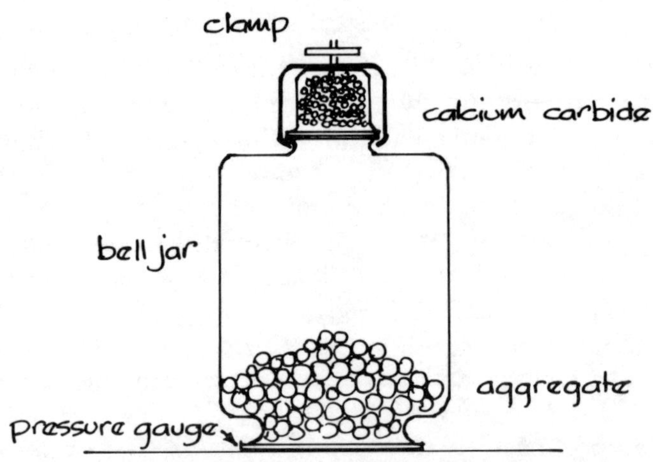

Fig 3.4 Moisture content determined by chemical reaction

Bulking

A subsidiary effect of moisture on sands (fine aggregate) is that a small amount of moisture will cause the sand to increase in volume (to bulk). This is caused by a film of water coating the surface of the particles and pushing them apart. However, when the sand is saturated, this film is broken down and the volume returns to its dry value. Because sands used in mortars and small-volume concreting work are often batched by volume, it is important to determine this bulking effect. The damp sand is placed in a glass cylinder and its depth is recorded. The cylinder is then topped up with water and the sand agitated so that it becomes saturated. The sand then settles and its depth can be remeasured. The apparatus is shown in fig. 3.5 and it can be seen that if V_b represents bulk volume and V_s the saturated volume, the percentage bulking equals $\frac{(V_b - V_s) \times 100}{V_s}$.

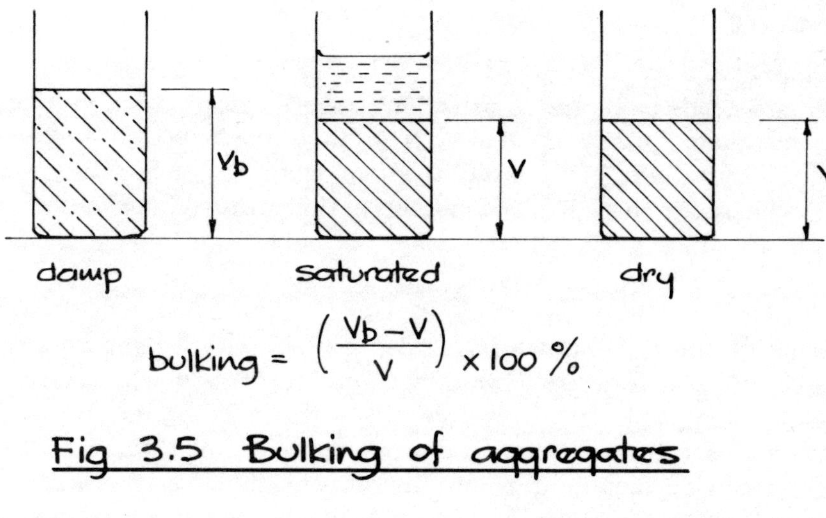

Fig 3.5 Bulking of aggregates

Cleanliness

Whereas it is easy to see whether large aggregates are clean, it is not so easy to notice silt in sands. However, a simple test can be carried out using a glass cylinder and water with a salt solution of 1%. This is poured on to the sample and the cylinder is shaken vigorously to mix the sand, salt and water. The sand settles quickly to the bottom of the cylinder but the silt is held in suspension and appears as a layer on top of the sand, as shown in fig. 3.6. If the silt layer is measured

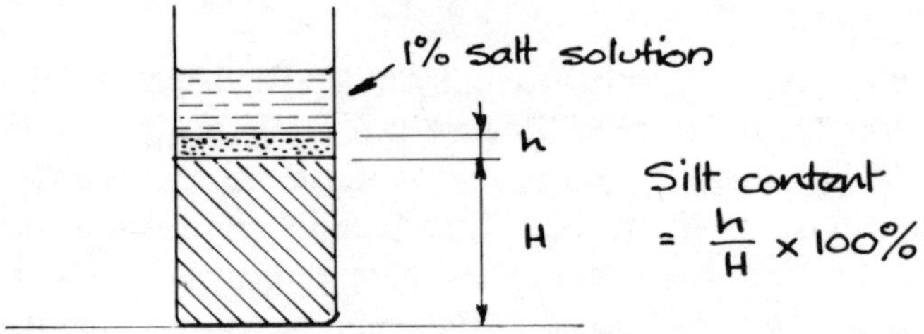

Fig 3.6 The silt test

as h and the sand layer as H, then $h \div H \times 100$ = percentage of silt. A limit of 8% silt content is acceptable, but if this is exceeded, the sand must be washed. Care should be taken in the washing process that all the fine grains are not removed.

CONCRETE

Workability

Concrete needs to be tested at various stages of its production to ensure adequate quality control. For freshly mixed concrete workability is important, being part of the specification for major structural work. There are three ways of measuring this properly, depending on the nature of the work:

Slump test

A cone of the type shown in fig. 3.7 is filled with a representative sample of wet concrete placed in four layers, and each layer is rodded twenty-five times. The mould is trowelled, carefully lifted clear, and set to one side. The wet concrete will slump and the amount can be measured against its cone height. If a true slump is not achieved, but a collapse or shear fault occurs, try again so that a sensible reading (to the nearest 5 mm) can be taken. This will give an indication of workability since a small slump will indicate low workability and a large slump high workability. It is not, however, very precise and is not considered sufficiently accurate for major contracts.

Compaction factor test

This gives a better indication of workability, particularly for mixes

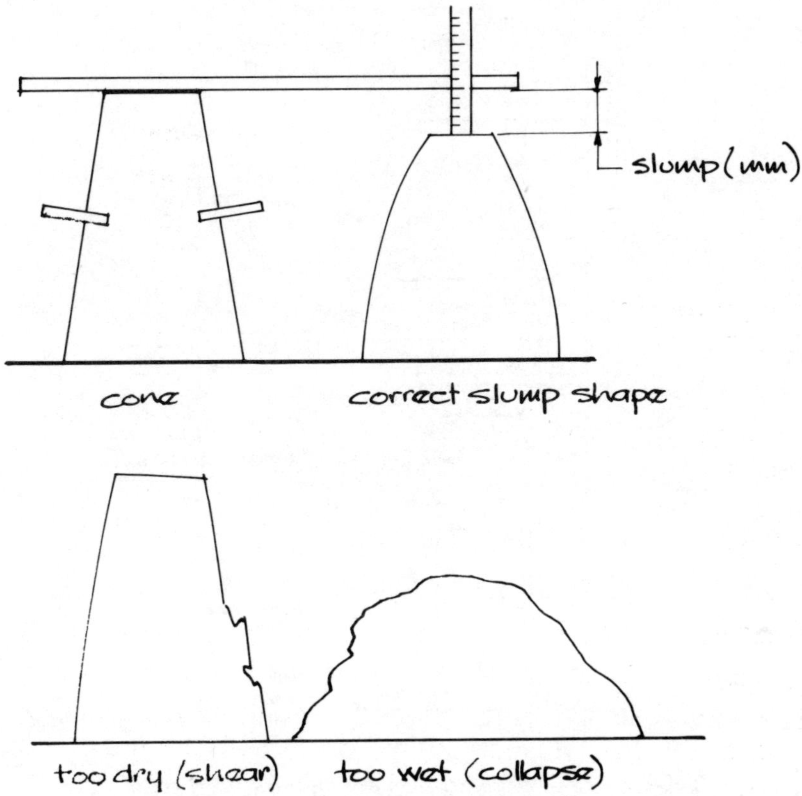

Fig 3.7 Slump test

with a low water/cement ratio. The apparatus shown in fig. 3.8 (in cross-section) is used, concrete being placed in the top hopper. The hinged flap is released and concrete drops into the lower hopper. The process is repeated with the concrete dropping into the cylinder placed below the second hopper. The surface of the cylinder is then trowelled off and the self-compacted concrete weighed. [*Note:* Since this includes the weight of the cylinders, its empty weight should be deducted.] This weight is recorded and the cylinder is emptied to be refilled with layers of concrete, each fully compacted. This fully compacted concrete is also trowelled off and weighed, the compaction factor being the weight of partly compacted concrete divided by the weight of fully compacted concrete. A high compaction factor will indicate high workability caused by a high water/cement ratio,

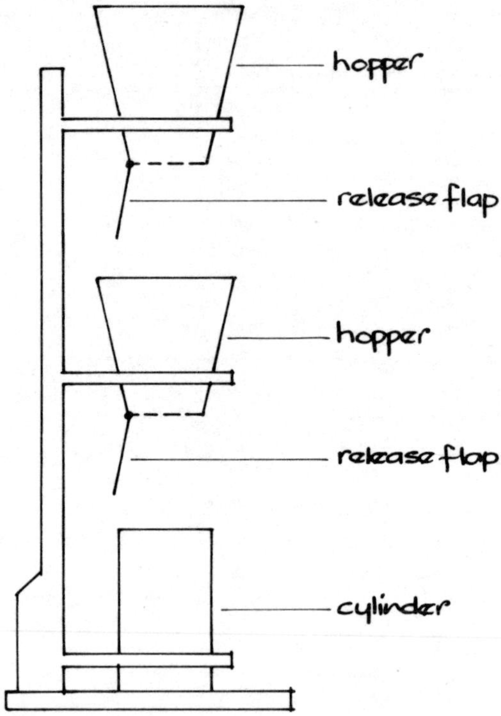

Fig 3.8 Compaction factor test

whereas a low factor will indicate low workability and a stiff, dry mix.

VeBe consistometer

As an alternative to using the compaction factor, the apparatus shown in fig. 3.9 may be used. A cone of concrete, using the mould from the slump test, is placed in a cylinder and vibrated until it takes the form of the cylinder. The transparent disc falls as the concrete is compacted and allows you to see when the shape has been re-formed. The time is measured in seconds (the VB time). This can also be used to specify workability for concrete mix design. Table 3.2 gives a comparative indication of the three methods, for a standard concrete mix with a maximum aggregate size of 20 mm.

Some concretes used on sites are pumped into position and need to have fairly high workability. Other methods are available to measure the flow properties of concretes used in such situations.

Table 3.2 Indication of concrete workability

	Specified workability Low	Medium	High
Slump (mm)	12–25	25–50	50–125
Compaction factor	0.82–0.88	0.88–0.94	0.94–0.97
VB time (s)	7–20	3–7	1–3

Strength

Hardened concretes develop strength and it is this strength that is of importance in structural concrete. Because the tensile strength of concrete is low, and is compensated by using steel reinforcement, it is the compressive strength that is of significance in determining the structural quality. By far the most common testing technique is to sample the concrete during placing by taking test cubes.

Cube testing

Cube testing involves casting, curing and storage of the sampled concrete in accordance with B.S. 1881 and should be carried out by trained personnel, possibly technicians. The cubes must be compacted so that they have a quality similar to that of the concrete placed. They are then vibrated for given locations. Also a set of cubes should be made for the batch of concrete so that tests can be carried out at different stages. The usual times are 7 days and 28 days, these being

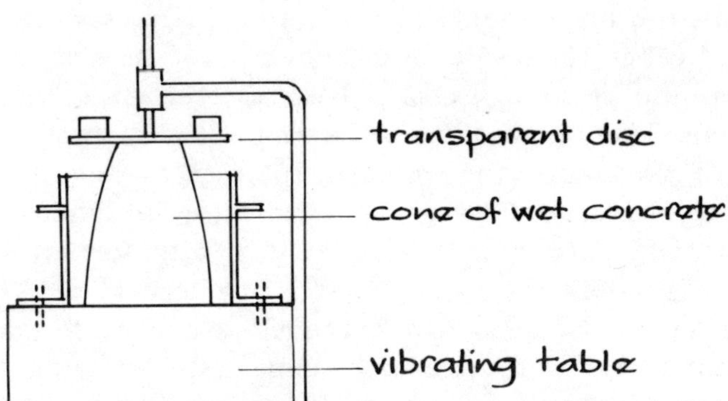

Fig 3.9 The VeBe consistometer test

required in the work specification, but 3-day cubes (and on critical sites even 1-day cubes) are often tested to give the contractor an early idea of the concrete strength as it develops.

Reference back to chapter 2 will remind readers that the hardening of concrete slows progressively, and if a target strength at 3 days is achieved, it is unlikely that the 7-day or 28-day test cubes will fail to meet their target strengths. The actual testing may be done on site or in a laboratory and comprise a compressive testing rig, which exerts a direct force on the cubes. The force is increased at a gradual rate until shear failure (see fig. 3.10) occurs. This is then recorded and translated into stress, based on the plan area of the cube, to give

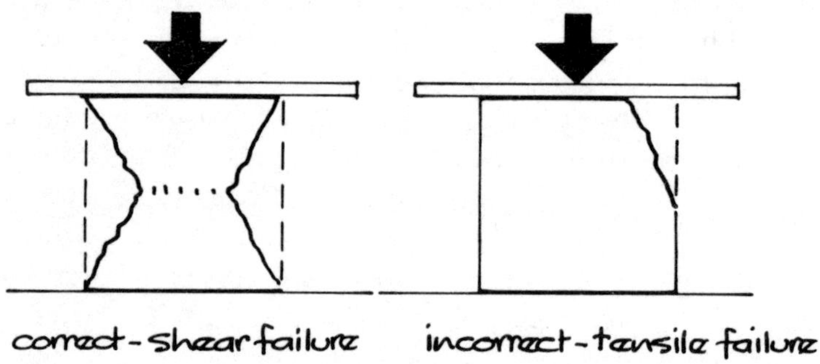

Fig 3.10 Cube crushing test

an ultimate stress in newtons per square millimetre. In chapter 2 we spoke of statistical distribution of results and it must be emphasised that not too much should be read into the figures obtained from one isolated cube. The results of a series of cube tests should be plotted and the use of control charts, based on statistical theory, is quite common. These generally give 'warning' and 'action' levels derived from a knowledge of the normal distribution.

In normal circumstances the contractor is unlikely to produce concrete of poor quality since test results are not known until perhaps a month after the concrete has been placed. If the mix is then rejected, it could prove extremely expensive to make good the deficient work. For this reason, cube tests are often much higher than required by specification. If, however, the mix does appear suspect, it is often prudent to make a further check on the strength of the concrete by taking core samples.

Core testing
Core testing involves cutting cylindrical samples of the actual structure, using diamond-tipped cutters, and the samples will usually be taken at relatively low stressed areas of slabs or walls. The core samples are then tested by direct compression in a similar way to the cubes described earlier. Core testing is only carried out in exceptional cases, either as a confirmative back-up to cube tests or where an existing building is to be adapted for different use.

Non-destructive techniques
A less dramatic but equally effective means of assessing the strength of hardened concrete, when cube tests may be at warning level, is to use non-destructive techniques. Two widely accepted techniques are the rebound hammer and the ultrasonic pulse velocity meter.

The rebound hammer is 'primed' by pushing it against the surface of the concrete. This triggers a mass which strikes the surface and rebounds, recording the magnitude of the rebound. This can then be translated to derive a concrete strength. Two precautions must be taken, however, if the results thus obtained are to be of any real value.

(a) The hammer must be at right angles to the surface to produce an accurate rebound result.
(b) The test location must be away from reinforcement, which could give an artificial reading.

The ultrasonic pulse velocity meter requires access to both faces of the concrete to be tested. A steel backing plate is located on one face of the section under test and a high-frequency sound is passed through the concrete. It rebounds from the plate and passes back, logging the time taken. This should remain consistent for tests over a wide area of the section, and knowing the E value of the material an idea of its strength can be derived. Its main benefit, however, is in detecting weak zones where honeycombing may be suspected or where cracking is thought to have occurred to some depth rather than just at the surface. In these situations the pulse velocity accelerates, showing a reduction in material density. If, however, the meter is located directly over reinforcement, the velocity will slow down and as with the rebound hammer an artificial result will be recorded.

It is therefore useful to know where the reinforcement is before carrying out either of these tests. It is also sometimes helpful to establish the location of reinforcement and its diameter and distance from the surface for other reasons. At one time this involved X-raying the section to highlight the reinforcement but a relatively cheap and simple method, since the reinforcement is made from ferrous metal, is to use electromagnetic methods. An electromagnetic device known as the *cover meter* can be used to locate reinforcement and then, having ascertained its location, can be adjusted to determine the distance of reinforcement from the face at which the meter is applied (in some meters it is also possible to determine the diameter of the reinforcement, though this is not generally necessary). Checking the 'cover' to the reinforcement is important since, if it is too near the surface, it may be prone to corrosion, and if it is too far away, it may not be developing its full 'lever arm' (see Volume 2.)

BRICKS

Some useful tests to ascertain properties which could affect selection for particular locations are now described.

Strength

Strength can be determined by filling the 'frog' of the brick with mortar (standard mix), placing it between plywood sheets (to distribute the applied force evenly) and then testing it for compression in the same way as for concrete cubes. This is only important when using 'calculated brickwork' since normal low-rise construction is unlikely to produce excessive loading on the brickwork.

Soluble salt content

It is useful to determine soluble salt content in order to assess whether the brickwork is likely to suffer from efflorescence or, more seriously, cryptoflorescence (as discussed in chapter 2). In B.S. 3921 is described a test that involves standing the brick in distilled water and allowing the water to permeate the brick by capillary action. As it evaporates, it draws the soluble salts to the surface and the extent to which they are present can easily be seen.

Water absorption

This involves determining the degree to which a brick absorbs moisture and may be a useful guide to its resistance to frost attack. A better guide, however, to assessing the weathering characteristics of particular types of brick is to actually build small 'test walls', which can be exposed to the elements over a period of time. These may be built at the builder's yard or at a research centre where they are available for inspection.

TIMBER

All structural timbers must now be graded so that they can be used with confidence in structural applications. This can be achieved in two ways, visual stress grading and mechanical stress grading.

Visual stress grading

Visual stress grading is primarily concerned with the visible defects and density of grain in the timber. A timber that is close grained, the grain running straight, and has no knots or twists, should perform better than a timber which has such faults. Knots are sections of branches and, if growing at the time the tree was felled, are described as 'live'. These should not seriously affect the strength of the timber, but if the branch was dead and a 'dead' knot is present, this may shrink away and fall out of the timber, producing a weak spot. Other faults are shown in fig. 3.11 and include 'shakes', 'wane', splits and fissures, all of which can weaken the section. Straightness of grain can easily be checked by scribing the section, i.e. inserting a scribing pin and pulling it along the section. It will naturally follow the grain of the timber and indicate its direction.

Mechanical stress grading

Much of the timber rejected using the techniques just described is not necessarily of poor quality and flexural loading tests have shown this to be true. Unlike manufactured materials, the strength cannot be determined from testing samples since every section of timber is individual in its strength characteristics. Mechanical stress grading overcomes this problem and is carried out by timber importers or large distributors on all timber that deserves the title 'structural grade'. The sections are passed, at speed, through a series of rollers that exert a load at 150 mm intervals, which deflects the section either up or down alternately. The load is plotted against the deflection electronically and fed into a computer which determines the section's elasticity. A splash of dye (colour coded) is sprayed on to the section at that point, and as the section finishes its run, the weakest record of elasticity is sprayed across the face of the section. Since elasticity is related to bending and deflection, the elasticity of the section will be an excellent guide to its permissible bending stress.

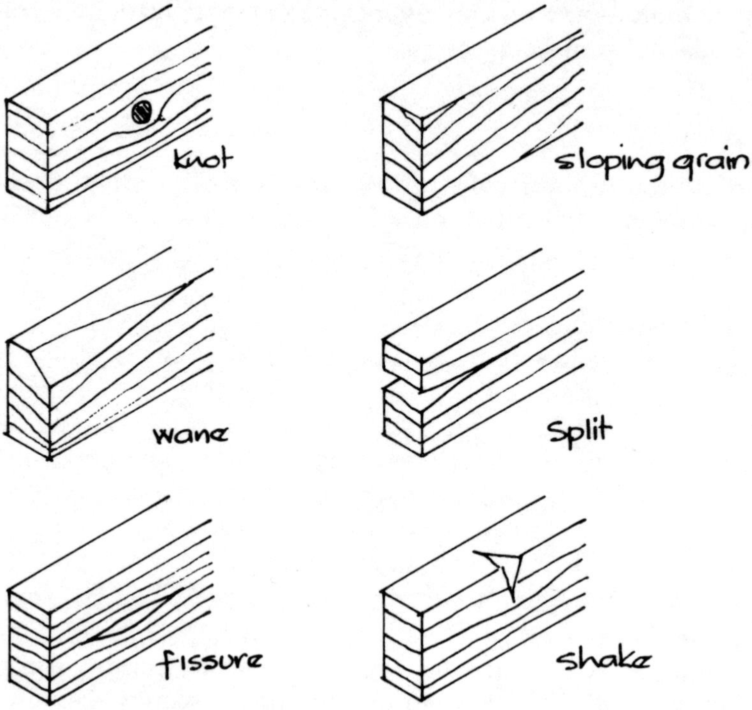

Fig 3.11 Common defects in timber

This method of grading has resulted in a much lower rejection rate, with a corresponding saving in costs.

Moisture content

An equally important aspect of timber testing is in determining its moisture content for the reasons outlined in chapter 2. This can be carried out prior to location or after the building is in use and involves the use of an *electrical resistance moisture meter*. This is a portable instrument, which has two probe needles. The moisture content is measured at the tips of these needles, giving a reading for a level below the surface of the timber. As a cautionary note the range over which the meter is accurate is between 6 and 24% moisture content (M.C.). Since all construction timber should have a moisture content within the limits of 6 and 22% (see table 3.3), this is acceptable as a useful method of quality control when it is suspected that timber may have been incorrectly stored or not protected during its exposure to the atmosphere.

METALS

The properties of metals that are of most interest constructionally are as follows:

(a) *Strength:* the ability to withstand large forces, whether in compression, tension, shear, bending or torsion.
(b) *Elasticity:* the ability to resist large forces, whether in compression, tension or deflection, without excessive deformation.
(c) *Ductility and malleability:* the ability to undergo a change of shape (cladding sheets, reinforcing bars).
(d) *Toughness:* the ability to withstand impact or suddenly applied loads.
(e) *Hardness:* resistance to indentation (useful, for example, in steels for tools).

Table 3.3 Moisture content range recommended by appropriate standards

Location/function	M.C. range and B.S./C.P. reference (6 8 10 12 14 16 18 20 22)
Ground floor joists	C.P. 112
Carcassing timbers	C.P. 112
Upper floor joists	C.P. 112
Roofing members	C.P. 112
Windows/roof lights	C.P. 153
Wood-wool slabbing	B.S. 2543 & 2824
Wood trim (second fixings)	B.S. 584
Flooring (intermittent heat)	C.P. 201
Internal joinery	B.S. 1186
Particle boarding	B.S. 3444
Flooring (underfloor central heating)	C.P. 201
	dry rot safety line

The first three properties can be assessed by performing tensile tests on the metal and correlating the results with the other properties. Figure 3.12 shows the set-up for a solid steel rod. A uniformly increasing tensile force is applied to the specimen, which is gripped around its perimeter by a pair of 'jaws' at each end. The test may also be carried out in an electrically heated furnace to predict the metal's behaviour at elevated temperatures, such as fires, where steel undergoes a serious loss of strength.

The majority of metals used for structural purposes show some ductility, and so can be stretched permanently before failure occurs. This is explained more fully in chapter 5 (example 5.4 and fig. 5.9). The shape of the specimen at failure is shown in fig. 3.13, this being typical for a ductile material, and is known as a 'cup and cone' fracture. As can be seen, an appreciable reduction in cross-section occurs, which accounts for the *apparent* loss of stress prior to failure. [*Note:* Compression tests cannot be carried out so easily since the metal will buckle before failing. This, and the importance of deflection, are covered in Volume 2.]

Structural members often have to withstand shock forces, either during erection or in subsequent use. It is important to safeguard against brittle failure in such instances (at joints or welds, for example) but simple tests of ductility cannot provide an accurate picture of likely behaviour.

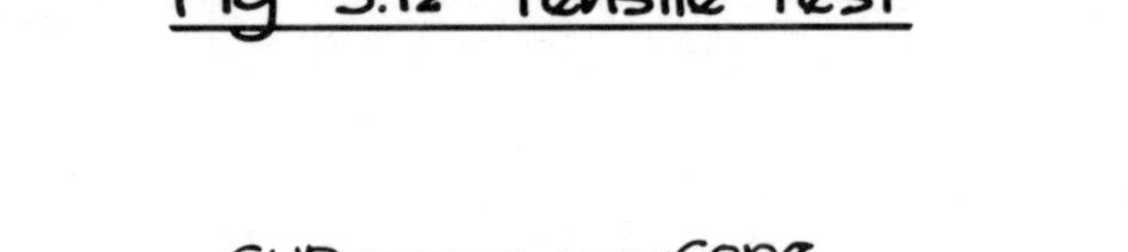

Fig 3.12 Tensile test

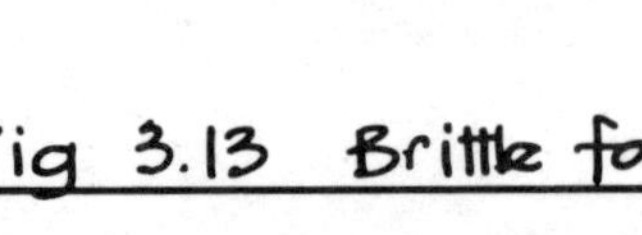

Fig 3.13 Brittle failure

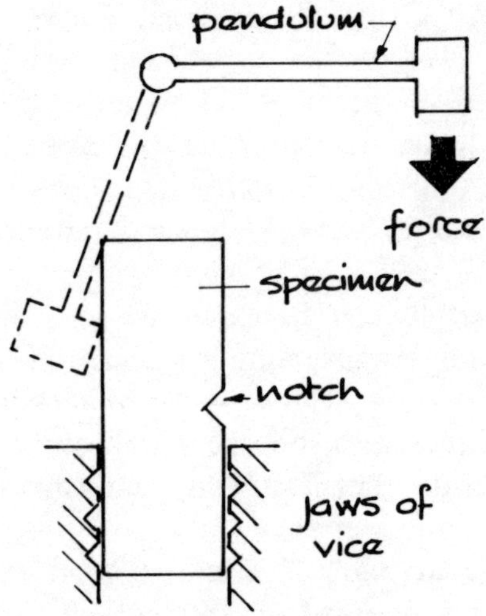

Fig 3.14 Impact test

A typical impact test rig is shown in fig. 3.14, where a loaded pendulum is allowed to swing from a fixed height to strike a specimen which has been notched to create an area of local weakness. In striking and *breaking* the specimen, the height to which the pendulum rises on the end of its swing is recorded. This indicates the measure of toughness and the impact energy absorbed is stated in joules. [*Note:* Testing is sometimes carried out at low temperatures where brittleness is likely to be increased.]

Hardness is measured by resistance to indentation. In the *Brinell hardness test* a steel ball (diameter 10 mm) is pressed on to the flat surface of the metal by a standard force, over a period of 10 to 15s, as shown in fig. 3.15. The diameter of the indentation is then measured by a mobile microscope, to determine the surface area of the

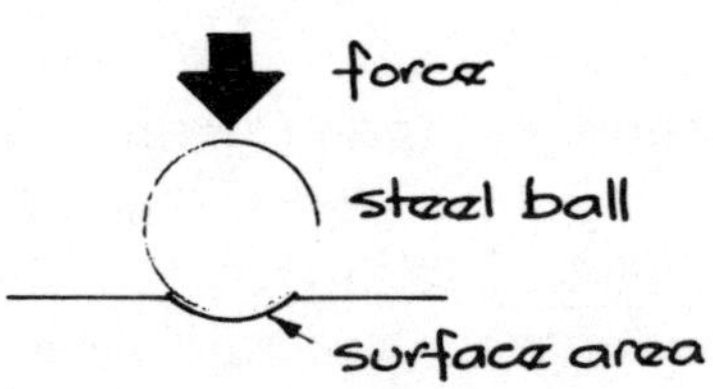

Fig 3.15 Brinell hardness test

indentation, and the Brinell hardness number (HB) is given by the applied force divided by the surface area of indentation. The applied force must not be too great or 'piling up' of the metal may occur. To achieve a sensible result a number of tests should be made and an *average* result taken. The Brinell scale is used for most metals but for very hard metals the *Vickers pyramid* indentation method may be used.

Metals may also behave in unfamiliar ways when subjected to prolonged, fluctuating or reversing stresses. *Creep* may occur under a constant high force, such as in a prestressing cable, and can be estimated by subjecting a specimen to a constant tensile force over a long period of time. The resulting extension in length can then be measured to determine creep.

Fatigue can occur under fluctuating load conditions where metals develop microscopic stress fractures caused by loss of ductility. They may then undergo a sudden brittle fracture. It is often possible to assess the number of 'stress cycles' needed to produce such failure. Figure 3.16 shows an apparatus that simulates alternating loads, where a metal rod is loaded at its centre causing deflection. The rod is then rotated at measured intervals through 180°, so that the compression and tension faces are reversed. However, wide variations occur between theory and practice and great care needs to be taken in interpreting the results of such a test.

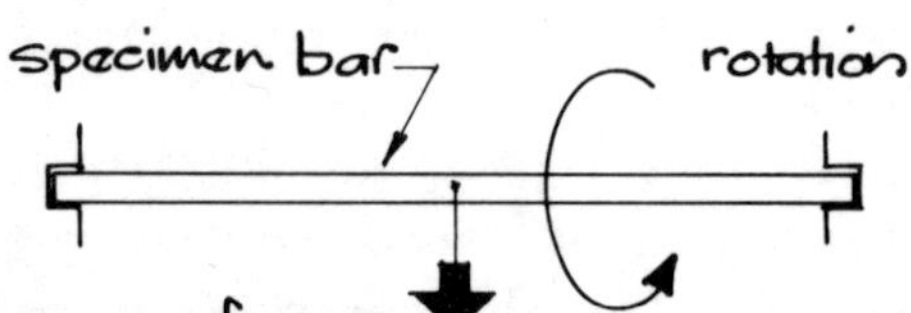

Fig 3.16 Fatigue testing

These tests are all carried out in ideal conditions and are intended as sampling tests since all are loaded to failure. Non-destructive testing on site is of more immediate interest to technicians and such tests may be necessary to monitor the structural behaviour of specialist structures subject to an aggressive environment, e.g. an oilrig platform, which is subjected to varying winds and wave forces in partnership with the corrosive properties of sea-water. This monitoring is intended to detect microscopic flaws, which act as 'stress concentrators' (see fig. 3.17) and can weaken the structure. As with reinforced concrete, X-ray techniques or ultrasonic pulse methods can be used. The latter method, however, does not involve time measurement (as in the case of concrete), since metals reflect the pulse. The method uses the fact that the reflected sound will be constant for a true surface.

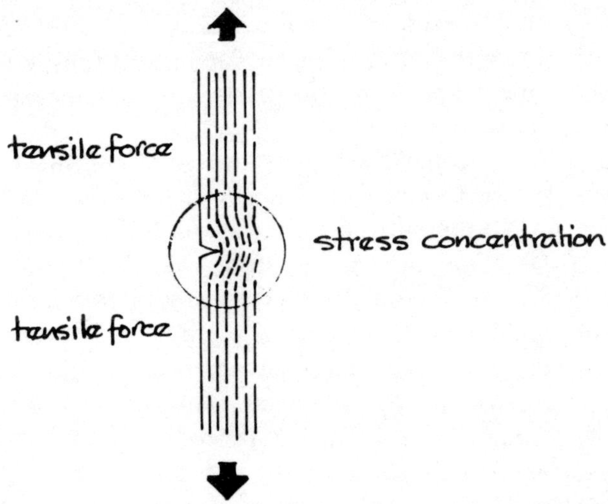

Fig 3.17 Effects of cracks on strength

PLASTICS, PLASTERS AND PAINTS

Since plastics, plasters and paints are not used structurally, tests tend to be concerned more with their useful life performance. Thus these materials tend not to be subjected to standard laboratory tests. Research establishments carry out a mixture of natural and accelerated exposure tests to assess performance and their published results are useful when considering the specification of these materials for a construction project.

BITUMENS AND TARS

Tests on bitumens and tars are classified according to whether they measure flow or composition.

Flow

Bitumens are graded by their resistance to surface penetration (as shown in fig. 3.18) using a standard needle of 1N force. This force is applied for 5s and the penetration is measured in *tenths* of a millimeter. To ensure consistency of test conditions, the test is carried out at 25°C. The grades are given in 'pens' (10 pens = 1 mm). As an example of material comparisons, mastic asphalt applied hot would have a penetration grade of, say, 20 pens whereas that applied cold, by trowel, would probably be about 400 pens.

The 'ring and ball' method (shown in fig. 3.19) measures a similar property, with a low softening temperature indicating a high penetration. The temperature at which the ball touches the bottom of the apparatus is the softening point.

However, the flow properties of some tars and 'cut-back' bitumens used in roadworks are too soft for the above tests. The 'standard tar viscometer' (fig. 3.20) measures the time taken for a specified volume of material to flow through a standard area hole.

Analysis of asphalts and macadams also requires a knowledge of composition in addition to flow properties and the composition tests involve chiefly the separation of the bitumen or tar binder from the filler/aggregate portion of the material. This may be achieved by 'extracting' the binder using a suitable petroleum-based volatile substance. On evaporation of the solvent, the binder is left and so its proportion in the original asphalt or macadam can be found. The materials remaining after the extraction of the binder can then be examined. If its 'grading' is important (this affects the amount of voids in the material, e.g. concrete), the material remaining may be sieved and its size distribution determined.

Assignment 3.1

Using the D.o.E. method, design a concrete mix to the following specification:

cement – O.P.C.
workability – 50 mm slump
aggregate – coarse (maximum size 20 mm) uncrushed; fine, zone 2, uncrushed
28-day cube strength – 35 N/mm^2

The minimum cement content is not specified so that an economic mix design can be produced.

What is the likely use for such a mix? Describe the quality control process that might be exercised in the production of the concrete.

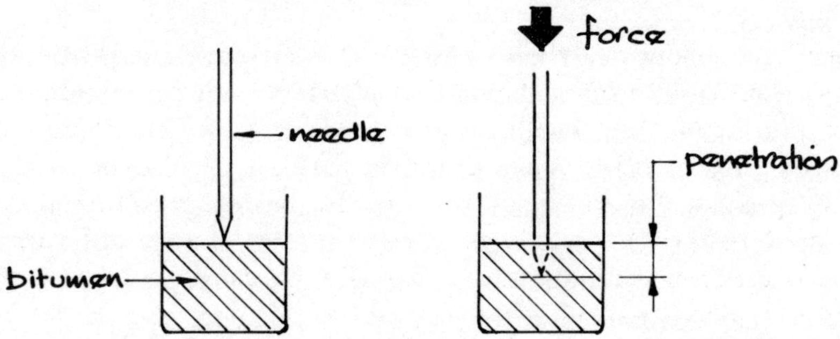

Fig 3.18 Penetration test for bitumen

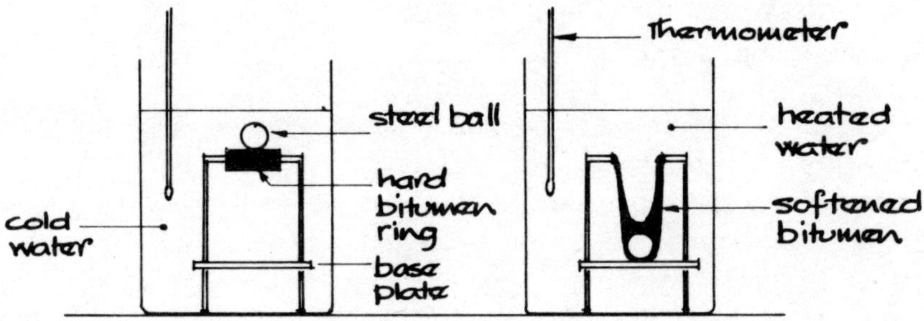

Fig 3.19 Softening point test

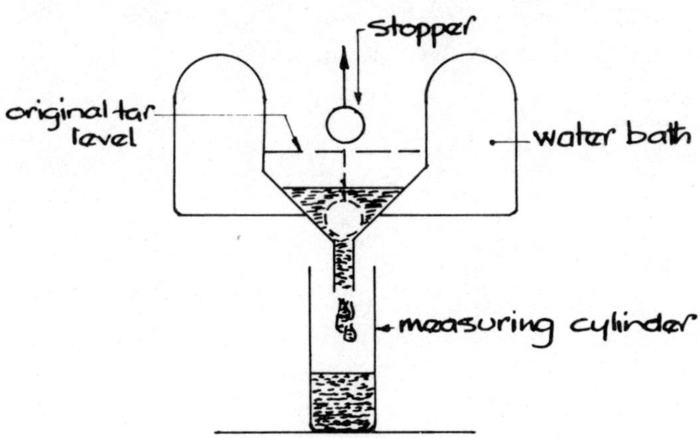

Fig 3.20 Standard tar viscometer

Assignment 3.2

Imagine that a new paint has been developed for the construction market. The manufacturer claims that it offers both protection and a pleasant-looking, long-lasting finish to steel. It is also claimed to need minimum surface preparation and may be applied in a single coat. An independent research laboratory is commissioned to evaluate the claims offered for the performance of the paint. Suggest how the manufacturer's claims might be investigated in a reasonably short time and then explain how the long-term characteristics might be determined.

II Mechanics

4 Applied Mechanics

In chapter 1 we talked about mechanics as being either 'applied' or 'structural'. The first of these categories is concerned with the *general* application of mechanical behaviour to a variety of topics including motion, energy, power and heat, and relates to much of the study covered in the environment. The second, however, is specific to the *structural* application of mechanical behaviour and how it affects the stability and strength of structural systems. This chapter, then, deals with general applications.

Mechanics in its basic sense is the study of motion or movement, what causes this and how it might be controlled. In order for anything to move, it must be subjected to some type of *force*. It is equally true that to stop something from moving it must also be subjected to a 'resisting' force. The most common type of force known is that of gravity. Waterfalls and landslides are good examples of gravitational movement, but the stability of objects is also created by gravitational force. Because of this we use the newton (N) to symbolise the unit of force around which we base calculations. One newton is the force required to accelerate a mass of one kilogram through one metre per second per second. Thus a falling stone accelerates towards the Earth, whereas a rocket requires a vast amount of force to escape (or exceed) the Earth's gravitational pull if launched into space.

DISTANCE–TIME DATA

Let us start with the concept of movement. If an object moves, it is often helpful to know how far it moves in a given time. In measuring distance travelled against time taken (known as distance–time data), we can establish *speed*, which may be measured either in metres per second (m/s) or kilometres per hour (km/h), depending on the types of distance concerned. Alternatively, if the distance is a constant based on circular movement, such as an engine or a record player, it may be expressed as so many revolutions per minute (r.p.m.). Some motors are required to 'run' at a steady rate of r.p.m., good examples being extractor fans for air changes, lift motors in tall buildings, pumps used in dewatering of trenches, and air-conditioning equipment. Others, such as lorries carrying excavated spoil, underground passenger trains, and auger drills for piling, can run at variable speeds to suit varying conditions. It is in this second category that we use distance–time data to establish *average* speeds.

Example 4.1

A lorry carries excavated material from a site to a tip 8 km away. The first 3 km are through busy traffic in a 50 km/h speed-limit zone and the time taken is 5 min. The remaining distance is covered in a further 6 min.

Plot the graph of distance against time to find:

(a) the average speed for the first 3 km;
(b) the average speed for the next 5 km;
(c) the overall average speed.

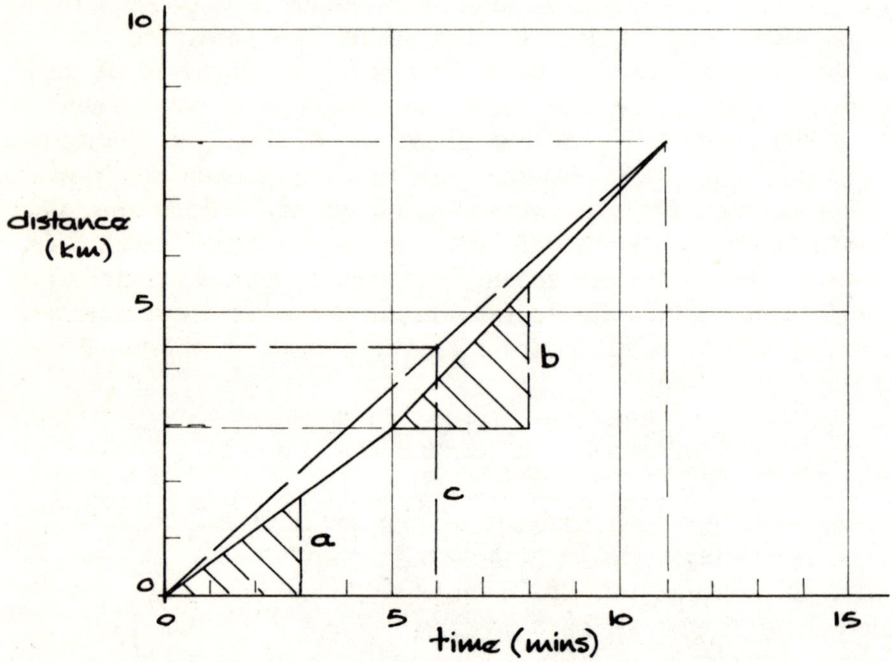

Fig 4.1 Distance/time graph

From fig. 4.1 we can see that gradient (a) represents the average speed of 3 km per 5 min, or, since we normally express this in hours, 3 ÷ 5 x 60 = 36 km/h. Similarly, (b) = 5 ÷ 6 x 60 = 50 km/h. Because the two gradients were each of straight-line form, we could, however, have plotted each over a distance of 3 min and multiplied by 20 to convert to hours, e.g. (a) 1.8 x 20 = 36 km/h, and (b) 2.5 x 20 = 50 km/h, which give us the same results. This is the technique we shall use for the overall average speed (c), shown by the dotted line gradient, since 11 min does not translate easily as a fraction of an hour. If we measure the gradient at the 6 min mark, we can plot the distance as approximately 4.4 km so the average overall speed is approximately 44km/h. [*Note:* It is unnecessary to be more precise than this.]

We can also use graphs of this nature to tell us what speed we need to achieve at any stage in a journey of given time limit. This can be very important for large-scale passenger transport, such as commuter trains, bus services and air-flights, but might be equally critical for site meetings using a car. Some 'executive' cars do have on-board computers, which can provide this information, but it can be seen on the distance–time graph as an interpolation against the overall journey. This is illustrated below.

Example 4.2

An architect leaves the car-park at his office at 9.15 a.m. to attend a meeting, scheduled to start at 11.00 a.m. at a site 90 km away. To take account of the time needed for parking and for a quick chat with the site agent, he allows 90 min for the overall journey. The route is made up of 10 km of urban roadways (office to motorway link), 72 km of motorway and 8 km of urban roadway (motorway link to site).

The first 10 km is covered in 25 min. The next 40 km takes 34 min, because of road maintenance, and an allowance of 15 min is made for the final stretch of urban roadway. From fig. 4.2 determine:

(a) at what speed the next 32 km of motorway must be covered;

(b) the average speed for the complete motorway section of the journey;

(c) the average speed for the first 10 km of urban roadway;

(d) the projected average speed for the final 8 km of urban roadway.

By plotting the distance–time data on the graph, we can see that

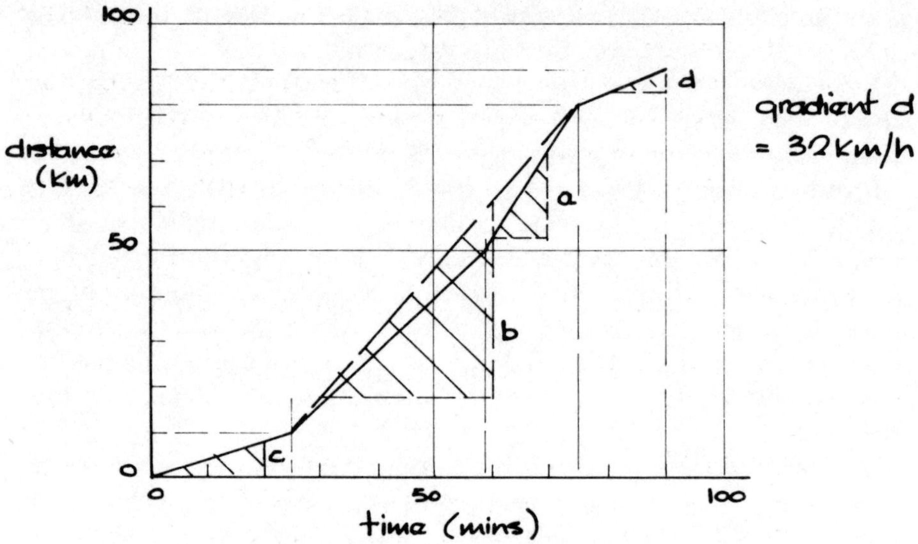

Fig 4.2 Distance/time graph

gradient (a) represents the steepest climb, and therefore the greatest speed, on the graph. If we interpolate between the 60 and 70 min points, we can measure a distance of 20 km, thus 20 ÷ 10 x 60 = 120 km/h. Similarly, by interpolating between the 30 and 60 min points on the dotted gradient (b), we can measure 43 km approximately. Thus 43 ÷ 30 x 60 = 86 km/h. Finally, by interpolating between the 0 and 20 min points on gradient (c), we can measure 8 km. Thus 8 ÷ 20 x 60 = 24 km/h.

In the example just considered, we were concerned with the actual distance travelled rather than the straight-line distance that might be measured on a map. If, however, we can relate the *time* taken (t) to a *straight* distance (s) we can determine *velocity* (v). Thus $v = \frac{s}{t}$.

VELOCITY–TIME DATA

In much the same way that we construct distance–time graphs, we can also construct velocity–time graphs. However, whereas distance–time gives us m/s (speed or velocity), velocity/time gives us m/s ÷ s = m/s^2, being the rate of *increase* in velocity, known as *acceleration*. A constant rate of acceleration will produce a straight-line graph showing gradient as acceleration. Figure 4.3 shows that this gradient can be dictated by either time or velocity, i.e. the shorter the time intervals, the greater the acceleration, or the greater the velocity increase, the greater the acceleration.

This may seem of little value when thinking in terms of cars, trains, etc., but it can be useful when dealing with electrical *impulses* or sound *waves*. Waves are caused by vibration and, when of a uniform nature (regular repeating shape), have certain regular features. These are:

wavelength – straight distance from start to finish of one wave
amplitude – height of peak or trough from the middle of the wave (pressure)
frequency – the number of waves generated per second
velocity – frequency x wavelength

It can be seen from this that, if the frequency is increased, the velocity is also increased. This can be put to good effect when testing concrete, or steel welds, for consistent density. If the velocity suddenly increases across a constant cross-section, it signifies a void or crack in the material.

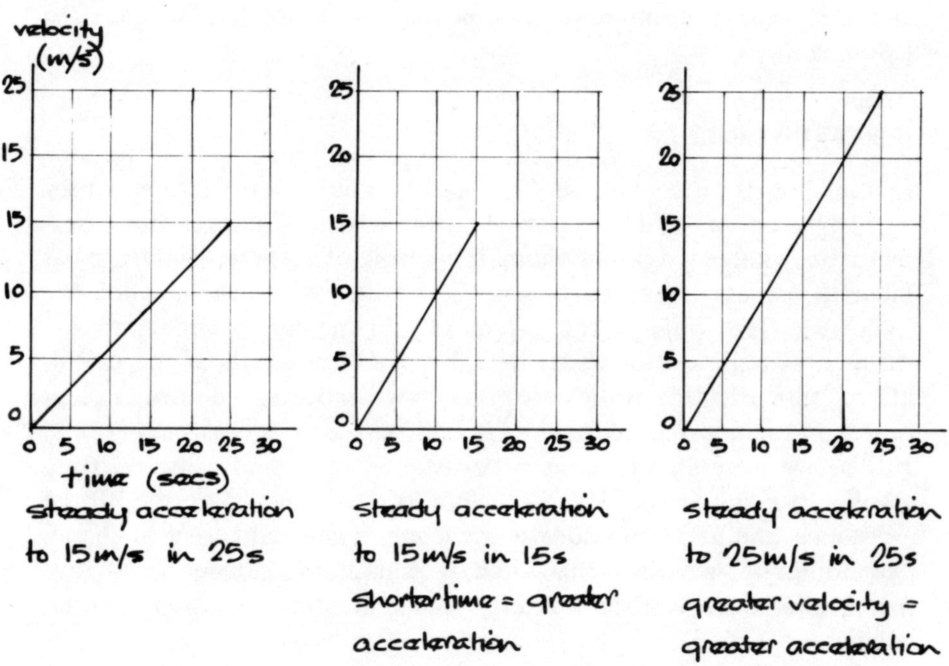

Fig 4.3 Velocity/time graphs

Waveforms are dealt with in greater detail in the chapters on electricity and sound but the basis of wave motion is shown in fig. 4.4.

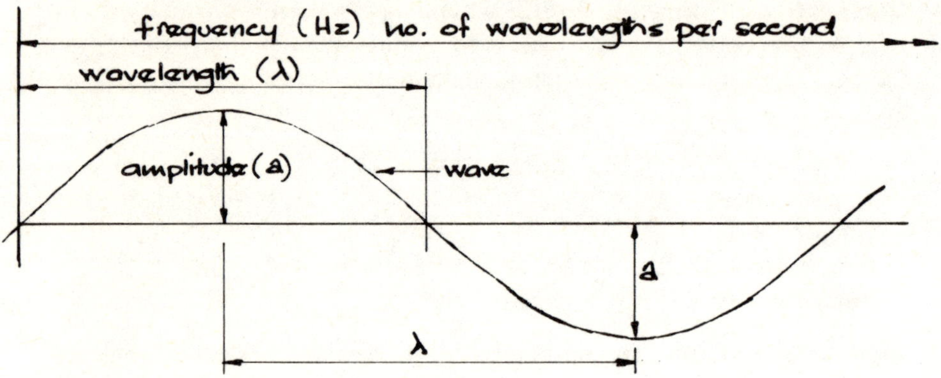

Fig 4.4 Wave properties

As can be seen from the diagram, the amplitude represents *pressure* and the wave motion represents *movement*, which has been caused by a *force*.

FORCE–DISTANCE DATA

In the discussion so far, we have talked about very large and very small movements and a variety of time scales. Whatever the movement, in whatever the medium, the result of a force displacing (or moving) a mass is known as *work*. The unit of work is called the joule (J). One joule is the effect of one newton (force) moving through one metre (distance) thus 1 J = 1 Nm. Much of the calculation applied in this book is concerned with joules, the prime exception being structural mechanics, as will be seen in the later chapters. Just as we were able to construct distance–time and velocity–time graphs, to measure speed and acceleration, we can also develop force–distance graphs to measure work done. These will vary in shape, depending on whether the force is consistent, subject to steady increase (or decrease), or variable. This is illustrated in the following examples.

Example 4.3
A batch of timber floor joists, requiring a lifting force of 4 kN, is to

be raised through a storey height of 2.5 m using a fork-lift truck. Determine the work done.

Because the force is consistent we can carry out a simple calculation, thus: 4 kN x 2.5 m = 10 kNm or 10 kJ. We could, however, plot the graph (fig. 4.5) showing the constant force as a horizontal line over the distance of 2.5 m. The area contained (a rectangle) is the work done.

Example 4.4

A prestressing cable of steel is to be stretched by 150 mm to induce a stress prior to concrete being cast around it. The tensile force required is increased steadily from an initial zero value to a final value of 8 kN. Determine the work done.

In this instance the area is triangular so that the calculation is slightly more difficult. Also the distance has to be translated into metres so that we calculate: 8 kN x 0.15 m ÷ 2 = 0.6 kJ, which may be better expressed as 600 J (see fig. 4.6).

Example 4.5

A run of steel sheet piling is rammed into the ground, against a resistance, to form the support for a proposed excavation. Because the ground is variable, being a mixture of sandy gravel and clay, and because of resistance to penetration by 'skin friction', the force required for each 200 mm penetration is variable. The following force–distance data are plotted on to the force–distance graph (fig. 4.7).

Force (kN)	0	4	5	5.5	6	7.5	9.5	12
Distance (mm)	0	200	400	600	800	1000	1200	1400

The graph can be seen to comprise a mixture of triangles and rectangles. Since all these have a common base of 200 mm, we can express the area as the summation of the triangular force ordinates (divided by 2) and the rectangular force ordinates, which is *then* multiplied by the distance of 200 mm, thus:

$$[(4 + 1 + 0.5 + 0.5 + 1.5 + 2 + 2.5) \div 2$$
$$+ (4 + 5 + 5.5 + 6 + 7.5 + 9.5)] \times 0.2$$

the forces being in kilonewtons and the distance in metres. This produces:

$$[(12) \div 2 + (37.5)] \times 0.2$$
$$= [6 + 37.5] \times 0.2 = 43.5 \times 0.2$$
$$= 8.7 \text{ kJ}$$

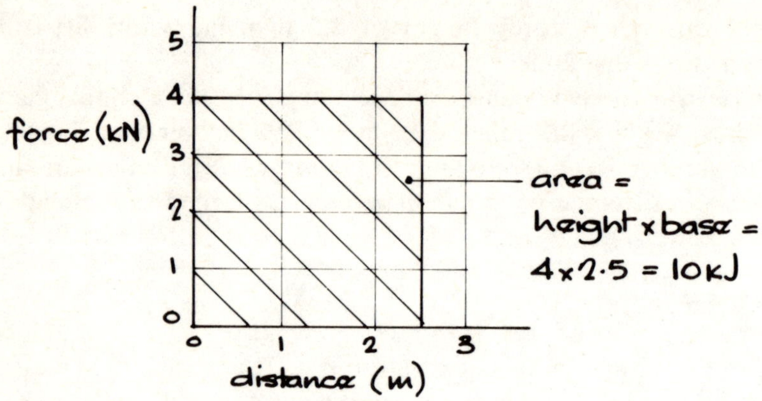

Fig 4.5

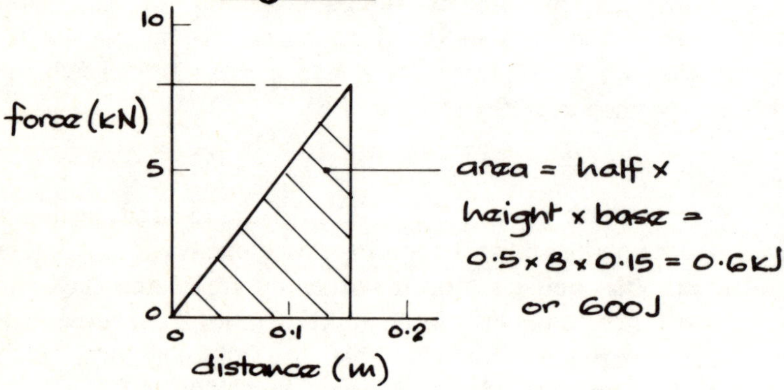

Fig 4.6

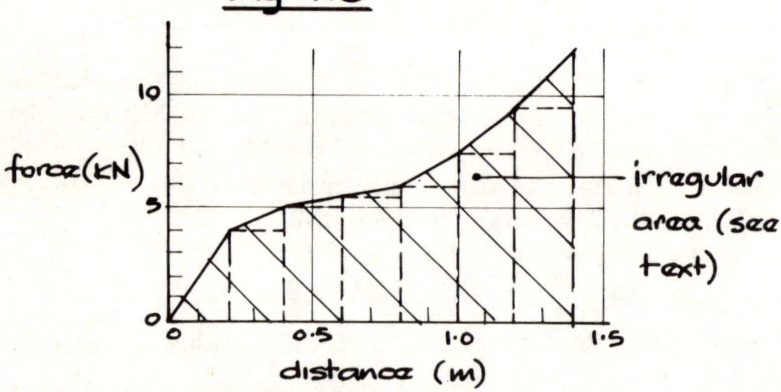

Fig 4.7

Force/distance graphs

ENERGY

Earlier in the chapter we talked about the relationship between speed and velocity. We can also see a special relationship between *work* and *energy*. In simple terms, energy is the capacity of something to *do* work. Without energy it is not possible to produce work. In scientific terms energy is never actually 'used up'; it is converted into a different form which may either be stored for future use or translated into a moving, or mechanical, energy. The two prime forms of energy are:

(a) *Potential energy:* This is the storing of energy for future use. Some simple examples of this are: coal, oil and gas (which store chemical energy in the elements of carbon and hydrogen); battery cells (which store electrical energy, which may be used for heat, light or magnetic movement); and the positioning of a mass (which may move to create gravitational energy).

(b) *Kinetic energy:* This is the movement of mass which, in meeting a body, causes work to be done. Some simple examples of this are: wind (which can drive windmills, move yachts or steer gliders); water (which can propel watermills, drive turbines to produce electricity, or provide the leisure pastime of surfing); or rotating cog wheels (which may propel a vehicle or drive a motor).

These then are the primary classifications of energy but the subdivisions are numerous. One of the most simple forms in which energy can be felt or seen is in the generation of heat. This may be caused by frictional resistance, by waste products or in some instances by the deliberate intention to provide warmth. If the last sentence sounds cynical, it is because this aspect of 'waste' energy has been so abused over the last century. Because the generation of heat has so often been regarded as a nuisance, and the abundance of fossil fuels such as gas, coal and oil (until the 1970s) has led to uneconomic use, it has never been thought of commerical benefit to harness the heat generated as a by-product of other energy expended. There have been a few notable exceptions, such as an occasional district heating scheme from a power station or a sewage plant, but in general the heat energy has been wasted.

Because of this, the energy input is invariably greater than the energy output of a machine or device. This can be expressed in terms of *efficiency*.

$$\frac{\text{Output}}{\text{Input}} \times 100 = \text{Percentage efficiency}$$

An example, even in this so-called energy-conscious era, is the gas-log fire, which incorporates gas jets lit to 'play' flames over ceramic logs to give the visual appearance of a real log fire. Although the visual effect is attractive, and may give a psychological feeling of comfort, the heat lost up the chimney compared with that radiated into the room or area in which the fire is located is very large. Some of these fires are only 28% efficient and very few exceed a 35% efficiency rating.

By far the most common reason for the production of waste energy in the form of unwanted heat is *friction*. This is the resistance of one material passing over another. Depending on the 'roughness' of the comparative surfaces and the gravitational influence, the movement will either be difficult or easy. The resistance to movement can sometimes be put to advantage such as in the resistance to sliding in the design of a retaining wall. Frictional resistance is also useful in arresting movement by 'braking'. The use of caliper, drum or disc brakes in stopping vehicles is commonplace but the greater the speed prior to braking, the greater the heat that will be produced. This has to be dissipated, and ventilated disc brakes are used on high-performance vehicles.

Unfortunately, frictional resistance is often an unwelcome component in the performance of machinery, and it needs to be reduced whenever possible by means of lubrication (by water, oil or wax), and by ensuring that surfaces are not unnecessarily roughened by either rusting or bluntness. Some simple examples of this are as follows:

(a) *Stone cutting:* a water jet is played over the cutting area to lubricate and cool the operation.

(b) *Hand saws for cutting timber:* should be free from rust, checked for correct tooth setting and sharpness, and either lightly oiled or waxed prior to use.

(c) *Masonry drills:* should be checked for sharpness and lightly oiled prior to use and should be turned slowly to reduce heat build-up from friction.

(d) *Cog wheels in a machine:* should be oiled lightly or coated with graphite to improve movement by reducing frictional resistance.

An alternative is to divorce the materials by the introduction of a secondary component. If it is required to move heavy objects over rough terrain, it might be useful to use *runners* or some form of *sledge* to allow the mass to slide. A more efficient device is the *wheel*. A common sight, in construction, is the wheelbarrow, which is used to transport a variety of materials and components. An almost equally common sight is the dumper truck, which is really something of a mechanical wheelbarrow, for work on a larger scale.

As mentioned previously, there is a relationship between work and energy and both are measured in joules (J). When we harness (control) energy, we produce *power*. In controlling the energy we can determine how much work might be done in a given time. Thus, power is the work done divided by the time taken. In controlling the work on a time scale, we have introduced another component and power is measured in joules per second. The unit of power is called the watt (W) and 1 W = 1 J/s. Because this unit is very small, we often use it, for convenience, in multiples of a thousand (kilowatts, kW) or even a million (megawatts, MW). For use with electronics, however, where we may wish to work in extremely small units of power, we might actually use divisions of a thousand (milliwatts, mW). In most cases in construction, we use kilowatts and as you will see in chapter 6, 'Thermal Studies', where we wish to cost the power used we sometimes work in 'kilowatt hours'. The relationship to the joule is therefore slightly complicated since we are changing from joules per second to joules per hour, which entails a multiplier of 60 x 60 = 3600. Thus: 1000 x 3600 J = 1 kWh (or 3.6×10^6 J = 1 kWh). This can be useful, for example, when costing the price of heating domestic water by means of an immersion heater of power output 3 kW. If the water is heated continuously for 1 h 40 min at full power, the cost would be based on a figure of 3 x 1.67 = 5 kWh. Many electrical appliances used domestically, such as wall fires, convector heaters, kettles and toasters, are very simple and convert electrical energy directly into heat energy. Because there are no moving parts and no waste, they are described as 100% efficient. Most appliances that produce power by energy conversion are not fully efficient and this needs to be taken into account when comparing running costs of different appliances producing the same power output.

Example 4.6
A central heating system, including a domestic hot water supply, needs a maximum output of 15 kW. Three boilers are considered.

(a) An oil-fired boiler requires an input of 25 kW because of waste combustion being discharged up the flue to roof level.
(b) A gas-fired boiler, with conventional flue, requires an input of 23 kW because of waste combustion being discharged as in (a).

(c) A gas-fired boiler, with balanced flue, requires an input of 19.5 kN since there is less waste combustion.

Determine the efficiency of the boilers and how this affects the cost of running them.

Boiler (a) has an efficiency of $\frac{15}{25} \times 100 = 60\%$

Boiler (b) has an efficiency of $\frac{15}{23} \times 100 = 65\%$

Boiler (c) has an efficiency of $\frac{15}{19.5} \times 100 = 77\%$

Because gas prices are consistent for boilers (b) and (c) we can ignore boiler (b). If for boiler (a), the cost of fuel per kilowatt hour is 4p whereas for boiler (c) the cost per kilowatt hour is only 2.8p, we can see that the true difference is *not* 4 ÷ 2.8 or 1.43 times the cost but (4 ÷ 0.6) ÷ (2.8 ÷ 0.77) because of the varying efficiency. This can be rewritten as (4 ÷ 0.6) x (0.77 ÷ 2.8) = 1.833, say 1.8 times the cost.

You may ask 'if all these machines are less than 100% efficient, how is it that we can raise loads, by using simple mechanical machines, that we are not able to lift directly?' Examples are levers (including crowbars) and pulley systems. A crowbar is often used to exert a much greater force than could be exerted directly by pushing or pulling against the object to be moved. Similarly, pulley systems are used to lift building materials, or even oneself, manually with less effort than by lifting directly.

What we must remember now is that work = force x distance. Thus, if we vary the distance so that the input force travels further than the output force, we should need less force to move the object than by direct application. When using a crowbar (or lever) we rotate around a fulcrum, with the output force close to the fulcrum and the input force a greater distance away (see fig. 4.8). You will see that for an output force of 1000 N acting at 100 mm from the fulcrum, an input of only 200 N is required if exerted at 500 mm from the fulcrum. In each case, however, the work is consistent at either 1000 N x 100 mm = 100 Nm = 100 J, or 200 N x 500 mm = 100 Nm = 100 J.

If we now look at a pulley system to a maintenance cradle, it will be seen to comprise two sets of three pulleys. This means that the pull on the rope has to travel 2 x 3 times the distance that the cradle

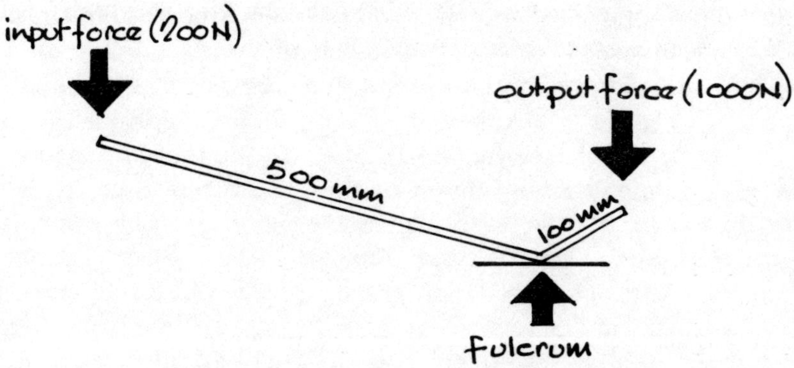

Fig 4.8 Lever mechanism

actually moves (see fig. 4.9). The force to be lifted is 2100 N but that exerted on each pulley is only 350 N. However, to raise or lower the cradle 5 m will still require a work effort of 2100 x 5 = 10 500 Nm or 10.5 kJ, half of which is applied at each pulley system.

The ability to move forces in this way is known as *mechanical advantage* and is used extensively in demolition and construction.

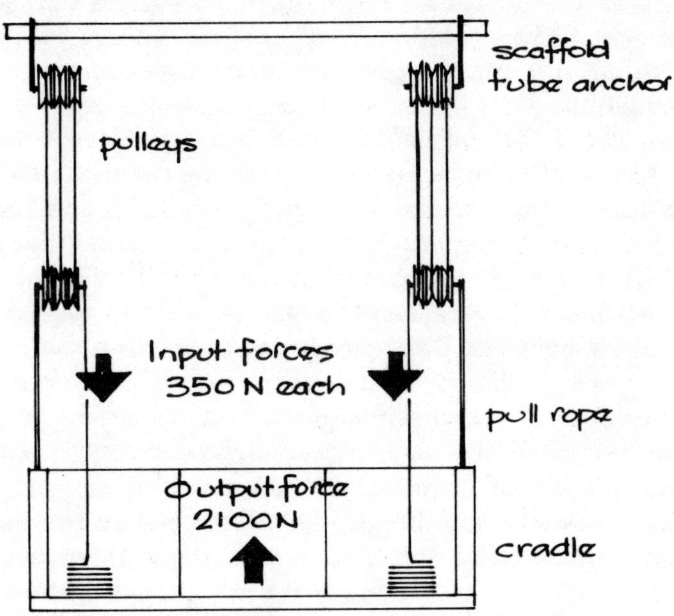

Fig 4.9 Pulley system

Further development of the principle of levers is covered in the next chapter, which deals with structural mechanics.

A third technique that can use mechanical advantage is the *screw*. A helical thread, interlocking with a 'mating' thread, produces movement when operated (see fig. 4.10). Because the travel of the thread is far in excess of the movement of the object, the force generated is also in excess of that used. There are many examples of this in construction, some typical examples being screw jacks, for underpinning; screw props, for temporary support work; screw vices, for carpentry work; and screw wedges, for fixings or, in extreme cases, for breaking up stone or concrete work. These are illustrated in fig. 4.11.

Returning our thoughts to heat energy, let us examine this energy form in a little more detail. We have already discussed the benefits and problems associated with friction and seen some of the more obvious ways in which heat can be utilised for the comfort of individuals. Although it is the most obviously recognisable medium in construction science, it should be realised that, as an energy form, it can cause movement, which may not always be readily appreciated. We also tend to talk about 'hot' and 'cold' conditions, associating heat with the first of these conditions only, whereas they are both different levels of heat within a range that we can sense quite easily. In order to measure the range, we use units of temperature. Prior to the introduction of S.I. units, the two commonly accepted temperature scales were Fahrenheit and Centigrade, the latter being based on the changes of state of water. Since metrication the Fahrenheit scale has fallen into disuse in scientific application and the Centigrade scale has been renamed as *Celsius*. An additional scale, *Kelvin*, which has the same graduations but starts at a different baseline, is used for scientific research, but as far as we are concerned, the Celsius scale is adequate. The 'changes of state' for water are: ice – solid, 0°C; water – liquid; steam – gas, 100°C. This means that in its most usual state, marginally above 0°C and below 100°C, water is a liquid. In solid form it may be many degrees below freezing point and where it takes the form of ground-water there are parts of the world, such as Alaska and Siberia, where it is always frozen. This is known as 'permafrost' and can cause problems in excavation. In other situations, in which ground-water in its liquid state is causing construction problems (such as the installation of a ventilation shaft or a tunnel), it may be useful to freeze it temporarily until construction work is complete.

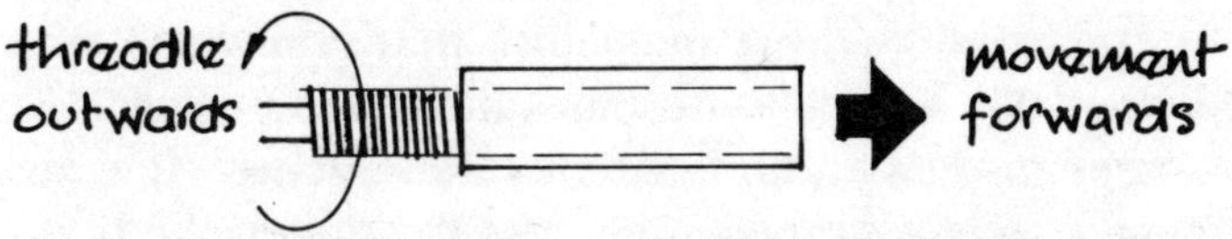

Fig 4.10 Screw thread mechanism

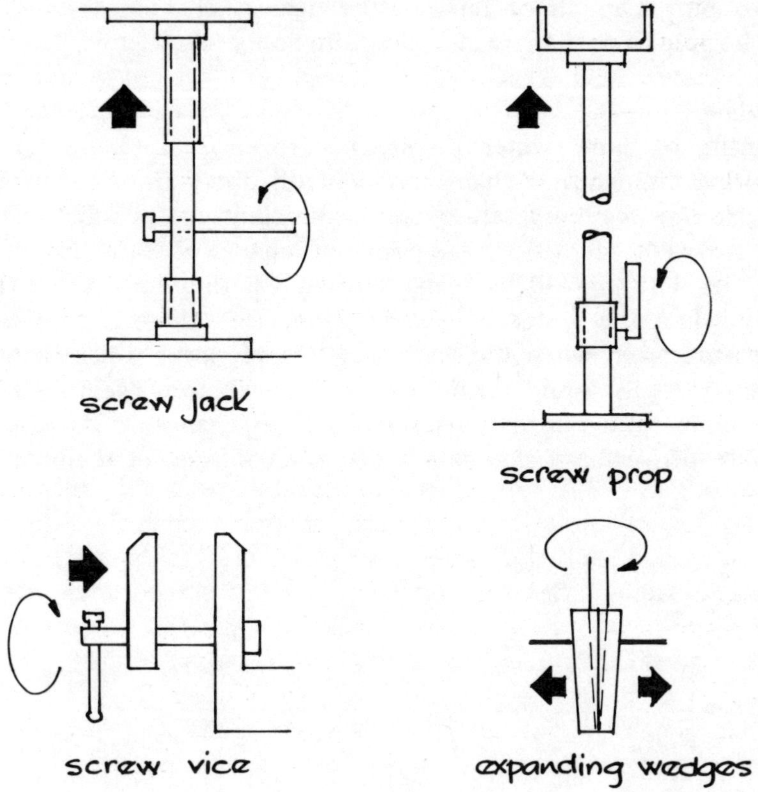

Fig 4.11 Constructional applications

In its gaseous state, water may be raised considerably above the 100°C point and in such conditions is known as 'superheated' steam. This is often utilised in heating systems or power stations. As you will know, steam is often used in the generation of power for engines such as steamrollers and steam trains. Let us consider what causes the changes of state and the movement (energy) involved.

When matter solidifies, the molecules that constitute that matter 'bunch up' together so that it becomes more dense. This is known as *fusion*. A rise in temperature agitates or 'excites' the molecules, which then start to move, becoming less densely packed, so that the matter becomes fluid (or liquid). If the temperature is raised sufficiently, the molecules are excited still further and can alter the structure from liquid to gas. This is known as *vaporisation*. If we were to plot a graph of temperature against change of state, we would be able to note these changes quite easily.

Example 4.7
A quantity of 'pure' water (distilled) is frozen prior to the start of monitoring temperature change. A controlled experiment is carried out registering the temperature rise as consistently as possible against a time scale and identifying the points of 'change of state'. From the graph (fig. 4.12) it will be seen that whereas the temperature rises steadily when the water is in solid form (ice), there is no rise in temperature from when the ice starts to melt until it has changed completely to its liquid form (water). Then there is again a steady rise in temperature until the water starts to change to its gaseous form (steam), when once again there is no change in temperature

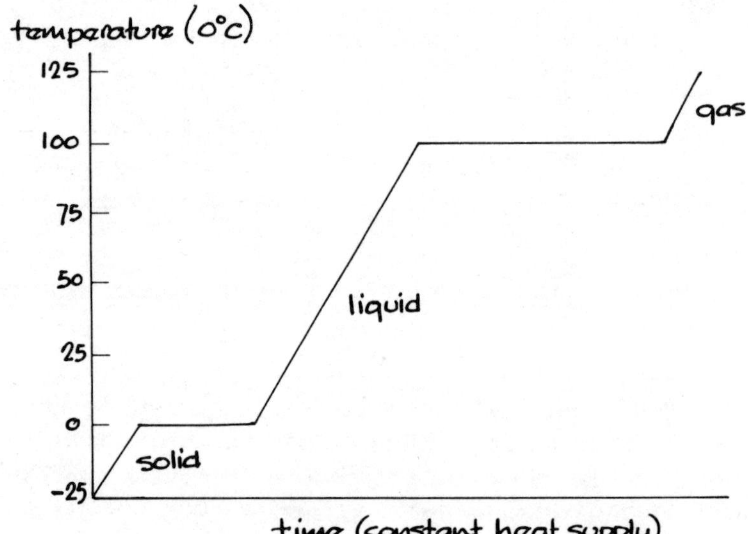

Fig 4.12 Temperature/time graph

until all the water has changed into steam. After this the temperature starts to rise steadily again.

The horizontal lines on the graph indicate to us not only the temperatures at 'change of state' as 0°C and 100°C, but also that a large amount of heat energy has been used at these points with *no* increase in temperature. Therefore there are two types of heat that have been registered:

(a) *sensible heat:* that which produces temperature change within the limits of a state;
(b) *latent heat:* that which produces the actual change of state without temperature change.

At the solid–liquid change, the latter is known as the 'latent heat of fusion', whereas at the liquid–gas change, it is known as the 'latent heat of vaporisation'. Table 4.1 gives the melting (solid to liquid) and boiling (liquid to gas) points for some commonly used construction materials. This gives an indication of the range of temperature difference between materials. Because of this it is possible to joint copper piping using soldered sleeves (a tin–lead solder is used) so that the solder is fully melted at less than 350°C with the copper unaffected. The table can also indicate why tungsten-tipped masonry drills are more efficient than iron-tipped drills, since the heat generated by friction does not affect the former so quickly.

Table 4.1 Melting and boiling points for some construction materials

Material	Melting point (°C)	Boiling point (°C)
Water	0	100
Lead	327	1750
Zinc	420	907
Aluminium	660	2400
Copper	1083	2580
Iron	1539	2900
Tungsten	3380	5700

Just as materials have different melting and boiling points, so they require different amounts of energy to heat them. If we consider the sensible heat of a material we can, by experiment, determine its *specific heat capacity*: that is, the amount of 'heat energy' required to raise the temperature of 1 kg of that material through 1°C. It is worth noting that the same amount of 'heat energy' is given off by the material in cooling through 1°C. Approximate values for some construction materials are given in table 4.2.

Table 4.2 Specific heat capacities of construction materials

Material	Specific heat capacity (J/kg°C)
Water	4200
Lead	130
Zinc	390
Aluminium	920
Copper	400
Iron	450
Brick	840
Concrete (dense)	920
Wood	1700

From table 4.2 it can be seen that metals require less heat energy than composite materials such as brick or concrete, and water requires substantially more energy than any of the other materials listed. It does also mean that water cooling down gives off far more energy than the other materials, so that it is quite useful as an energy store, e.g. in central heating radiator systems. It is also worth noting that if contained in a well-insulated hot water cylinder, it can retain its heat overnight. Water heated by off-peak 'white meter' electricity can be stored in this way. If we look at the power used and estimate its approximate cost, we can see the value of this method of water heating.

Example 4.8

A hot water cylinder of 0.22 m^3 capacity has an initial water temperature of 10°C. It has to be raised to 75°C. Determine:

(a) the energy required (density of water 1000 kg/m^3);
(b) the power used (1 kWh = 3.6×10^6 J);
(c) the cost for off-peak electricity (say, 3p/kWh);
(d) the saving against normal price electricity (say, 5.94p/kWh).

(a) Translate volume into mass for the water = 220 kg and multiply specific heat capacity x mass x temperature range, thus: Energy = 4200 x 220 x 65 = 60.06 MJ.
(b) Translate energy into power, thus: Power = 60.06 ÷ 3.6 = 16.683 kWh.
(c) Cost = 16.683 x 3 ≏ 50p.
(d) Full-price electricity costs 16.683 x 5.94 ≏ 99p so that a saving of 49p is effected. (Since this is based on a nightly cost this means a saving of around £3.50 per week!)

One of the problems associated with the grouping or dispersing of molecules by temperature variation is that the volume of the mass is changed. In general, as a material increases in temperature, it increases in volume (expands) and conversely, as it reduces in temperature, it reduces in volume (contracts). The exception to this rule is water, which expands as it reduces from 4°C through freezing into ice (above 4°C it follows the general rules of expansion and contraction but on an accelerating scale up to its boiling point).

As far as solids are concerned, our main interest is their change in length (or any linear dimension such as height or breadth). These are very small when considering the movement per degree and relating such movement to a length of 1 m but if the temperature range is considerable and the length of material excessive, the movement can be quite large. Table 4.3 gives the coefficient of linear expansion for some common materials.

Table 4.3 Coefficient of linear expansion of construction materials

Material	Coefficient of linear expansion per °C
Lead	29×10^{-6}
Zinc	30×10^{-6}
Aluminium	23×10^{-6}
Copper	17×10^{-6}
Iron (steel)	11×10^{-6}
Brick	7×10^{-6}
Concrete (dense)	11×10^{-6}
Wood (along grain)	5×10^{-6}

You will notice from table 4.3 that steel and dense concrete both have the same coefficient of linear expansion. This is fortunate, since they are combined to form reinforced concrete. If their rates of expansion were dissimilar, with one material expanding or contracting more quickly than the other, it would cause distortion, which would not be acceptable for stability. This distortion, when using dissimilar materials which are joined together, can be used to advantage in certain conditions, the most common example being the bimetallic strip. If two metals of differing rate of expansion (say aluminium and steel) are fixed at intervals along their length, they will expand differently when heated. The aluminium will expand more than the steel but since it is fixed to the latter, it will have to curve around the steel, causing distortion (see fig. 4.13). When the temperature starts to drop, the strip will begin to straighten out again. By fixing one end of the strip and making contact points at the other end, we can see that on curving away the contact is broken, but on straightening out, contact is re-established. In introducing this into an electrical circuit, we have formed a thermostatic control which operates a cut-out when required temperatures are reached. This can be quite useful in domestic appliances, such as coffee filter machines, steam irons and ovens, but can also be utilised very effectively in central heating systems (room thermostats), autoclaves for producing building blocks, or storage sheds for woodblock flooring (to avoid swelling or shrinkage).

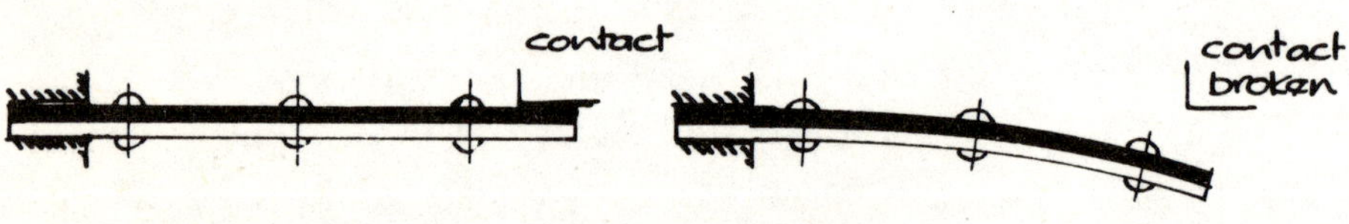

Fig 4.13 Bi-metallic strip

If no allowance is made for the thermal movement of materials over large distances, such movement can produce *strain energy*, causing either tension or compression in the material, which is unwelcome. Let us examine the effect on a concrete road surface.

Example 4.9

A concrete road with light mesh steel reinforcement is to be provided with expansion joints at 15 m intervals. The joints are formed at an air temperature of 5°C and should allow for a maximum air temperature of 38°C. Determine:

(a) the expansion between 5 and 38°C;
(b) the width of joint allowing a 10% safety factor.

(a) The expansion is easily determined by multiplying coefficient of linear expansion x temperature range x length thus: $11 \times 10^{-6} \times 33 \times 15 \times 10^{3}$ mm = $11 \times 10^{-3} \times 33 \times 15$ = 5.445 mm.

(b) Width required = $5.445 \times 1.1 \triangleq 6$ mm. In practice, the actual gap between the concrete bays will be greater than this, since the 'filler' material, normally a polysulphide sealant, has to be capable of elastic movement of 6 mm. This would mean that for an elastic distortion of 40% the actual joint would be 15 mm wide (i.e. $15 \times 0.4 = 6$ mm) (see fig. 4.14).

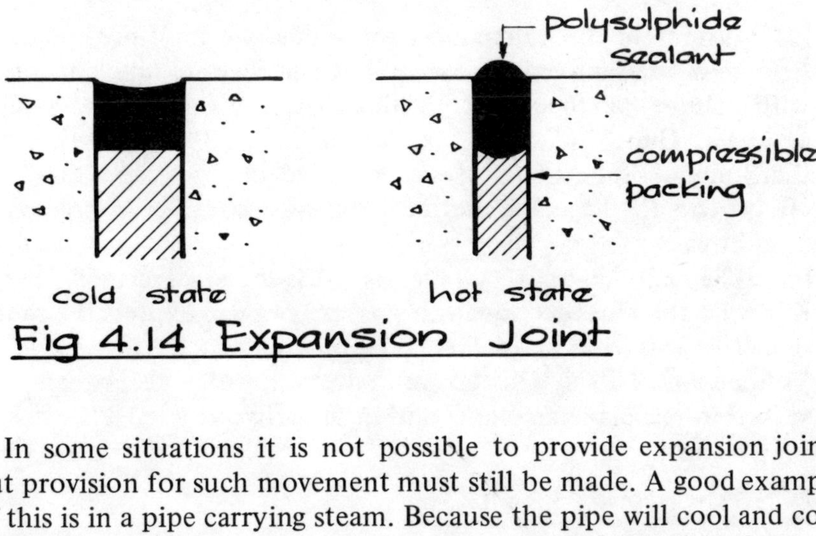

Fig 4.14 Expansion Joint

In some situations it is not possible to provide expansion joints but provision for such movement must still be made. A good example of this is in a pipe carrying steam. Because the pipe will cool and contract when not in use and may expand at 'superheated' steam temperatures, an expansion 'loop' is built into the system to take up this thermal movement.

Having discussed solids, let us look briefly at liquids. Because liquids do not have shape, they will expand in volume only, when treated, and will move to take up any space available within the container in which they are housed. If there is not sufficient space available, they will exert force on the container and may burst it.

Radiator central heating systems operate by using very hot water, which will expand. So that this expansion can be contained, a 'header' tank is provided. This will accept the increased volume and, when the water cools down, return it to the system. A hot water cylinder, if water is not drawn off for use, will direct the increased volume up an expansion pipe which may, in extreme cases, discharge into the cold water storage tank. Let us look at the effect for the hot water cylinder considered earlier, in example 4.8, to see the volumetric increase.

Example 4.10
A hot water cylinder of 0.22m^3 capacity contains water within a temperature range of 10 to 75°C. Determine:

(a) the volumetric increase over the temperature range (average coefficient of cubic expansion = 3.87×10^{-4});
(b) the length of expansion pipe of internal diameter 20 mm to take this expansion.

(a) Just as in the calculation for solids, we multiply the coefficient by the temperature range, but since we are now concerned with volume we then multiply this figure by the original volume of water. Thus: $3.87 \times 10^{-4} \times 65 \times 0.22 = 5.53 \times 10^{-3}\ \text{m}^3$. Since the volume of liquid is often expressed in litres, and there are 1000 litres to the cubic metre, the answer could be expressed as 5.53 litres.
(b) The volume of a cylinder is cross-sectional area x length. Knowing the cross-sectional area as πr^2, we can express the radius in metres as 0.01 m ($10 \div 1000$). Thus: Area $= \pi \times 0.01^2 = \pi \times 10^{-4}$. Volume $= \pi \times 10^{-4} \times$ length in metres (l), which also equals $5.53 \times 10^{-3}\ \text{m}^3$. Balancing the equation we arrive at $\pi \times 10^{-4}\ l = 5.53 \times 10^{-3}$, so that:

$$l = \frac{5.53 \times 10^{-3}}{\pi \times 10^{-4}} = \frac{55.3}{\pi} = 17.7\ \text{m}$$

From this it can be seen that two courses of action need to be taken:

(i) locate the hot water cylinder at ground floor level to allow a maximum head of pipe;
(ii) increase the diameter of the pipe to an internal dimension 30 mm so that an increase in radius of 50% (30 : 20 as 1.5 : 1) will produce an increase in area of 125% ($1.5^2 : 1^2$ as 2.25 : 1). This will reduce the length of expansion pipe required to $17.6 \div 2.25 = 7.8$ m.

It is also worth noting at this point that an additional benefit of the thermostat, apart from the economic consideration, is that it prevents all the water being converted to steam. If the *supply* of water were suddenly cut off, possibly during maintenance work, steam being heated would make the cylinder *implode* since the steam would escape via the expansion pipe.

Although you will have been told at secondary school about the ways in which heat is transmitted, let us look at the principles of conduction, convection and radiation in relation to some of the applications studied in this chapter.

Conduction allows heat to be transmitted from one material to another. This means that by heating an electric element (the immersion heater) we can heat the water around it.

Convection is caused by the material expanding when heated, thus reducing its density, so that if there is hot and cold water in a cylinder the hot water will rise to the top, allowing the cold water to be heated by the element. [*Note:* For economic use of electricity a 'two-bayonet' element may be used with an economy switch operating the short element only. Water below this element will not then be heated except by very slight conduction.]

Radiation of the hot water in a central heating system will be provided by the aptly named radiators to produce direct heat. The air will also pass around the radiators to produce convection currents.

The application of these basic principles will be developed more fully in chapter 6 on thermal studies but to impress you with regard to the 'power' of heat as an energy source it might be interesting to cast our minds back to the beginning of this chapter. We know that a rocket needs huge amounts of force to lift it clear of the gravitational pull of the earth, yet a hot-air balloon can reach to distances of approximately 24 km above the Earth's surface (almost beyond the gravitational pull) using only heat as its means of propulsion.

Exercises

4.1 A commuter train travels from station 'a' to station 'g', stopping at five intermediate stations en route. The timetable allows for a 90s stop at each station ('b' to 'f') to let passengers on or off the train. In table 4.4 are given the distance between each station and the arrival times at each station, for a starting time of 07.40. Determine, using a distance–time graph:

(a) the maximum speed of the train;
(b) the average speed of the train.

Table 4.4 Distances and arrival times in exercise 4.1

Station	Distance between stations (km)	Arrival time
a	0	*N*
b	2.0	07.43
c	2.5	07.47
d	2.0	07.51
e	3.5	07.55
f	8.5	08.06
g	2.5	08.10

4.2 A dumper truck transports a load of wet concrete (50 kN) a distance of 6 m at which point it deposits 30 kN of the concrete. It then travels a further 3 m and dumps the remainder. Calculate the work done during transportation of the concrete.

4.3 Explain the differences between potential and kinetic energy and give four examples of each, related to construction.

4.4 Explain the difference between mechanical efficiency and mechanical advantage, illustrating your reasoning by examples applied to construction.

4.5 A 'night storage' heater, containing dense concrete blocks of volume 0.16 m^3, has an initial temperature of 10°C. The temperature is raised, using off-peak electricity, to 40°C. Assuming a concrete density of 2400 kg/m^3, and using information contained within this chapter, determine:

(a) the energy required;
(b) the power used;
(c) the cost of heating the concrete.

Assignment
Examine the conditions in an office building that may create contraction and expansion problems, particularly with regard to materials used, and discuss how they are overcome in practice. Discuss the ways in which friction, associated with such thermal movement, can be minimised. Produce sketch details for movement joints with weathering provision as required.

5 Structural Mechanics

This chapter is devoted to the structural aspects of mechanics: the effects that forces have on the 'structure', the ways in which they are translated through the 'structural components', and the ways in which they may be resolved. The chapter will also provide an introduction to the ways in which the materials behave under such forces. Structural mechanics is based on a few straightforward rules and requires a certain amount of logic. Much of the work *can* be done graphically but the quality of draughtsmanship required is demanding in its accuracy. Do not be frightened off by thinking that a high level of mathematics is required, since this is *not* true. It is possible to use calculus, but it is unnecessary to do so, most calculations requiring only a fairly basic level of arithmetic.

EFFECTS OF FORCES ON STRUCTURES

Although we live in a three-dimensional world, a great deal of structural behaviour is considered in *two dimensions*, linked by the third. A simple example of this is a roof truss, which is a two-dimensional framework carrying loads from, and translating them back to, the third dimension (see fig. 5.1). There are two main reasons for this approach:

(a) Most of the early work on structural analysis was done graphically using geometry and scale measure. This necessitated thinking in *one* plane so that all work had to be considered about *two* axes.

(b) It is far easier to manufacture components, such as frameworks, in two dimensions than three and they are also easier to transport. Examples are portal frames, latticed beams and stud walling.

This does not mean that all structures are designed this way but as far as this level of work is concerned, we need only consider behaviour

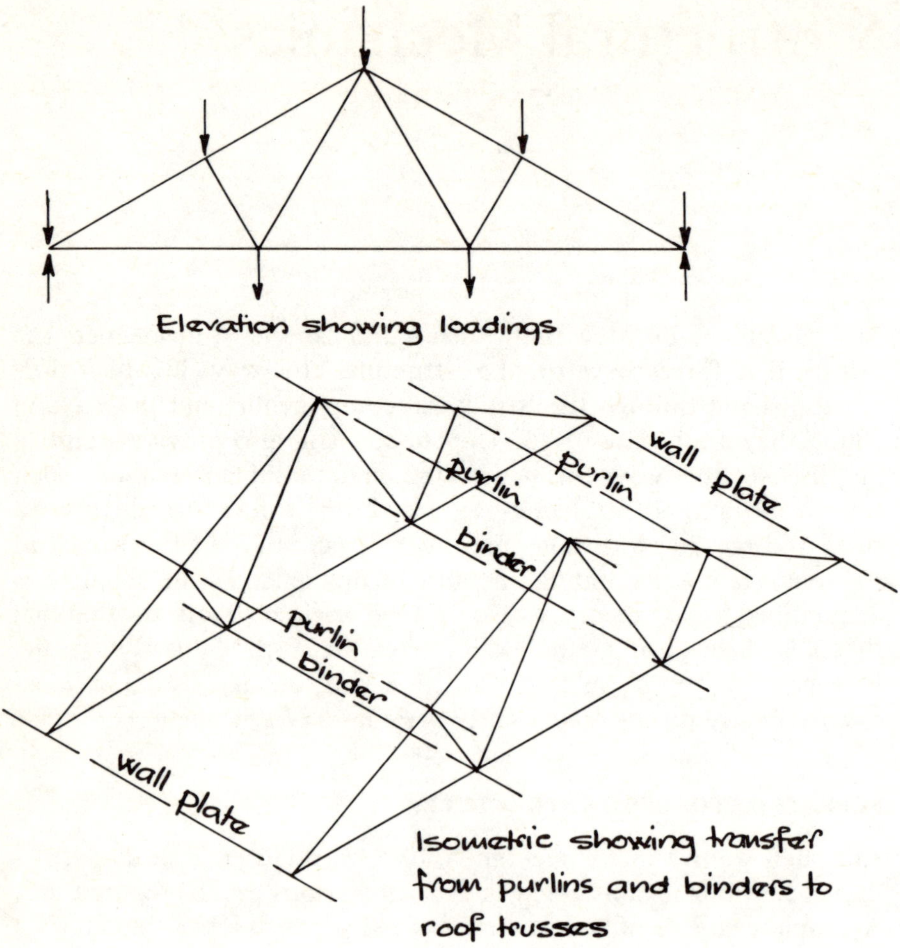

Fig 5.1 Showing transfer of loads from three dimensional to two dimensional systems

on one plane. In chapter 4 we talked about *straight* distances when considering velocity. Similarly, we think in terms of straight lines when considering forces. When a force is applied, or released, it will act in the most direct way until it is diverted into a different direction. When dealing with forces acting in one plane, this means that we can make the statement: 'All forces act in straight lines', and, since we are considering only one plane, they are known as *co-planar* forces.

They are also known scientifically as *vector quantities*. Do not be confused by this: it is an example, as described in chapter 1, of dressing things up in scientific language. What you need to remember is that:

(a) the forces can be plotted on a flat sheet of drawing paper (co-planar);
(b) the forces are represented by straight lines of a measured length (vectors).

In many instances, two or more forces may act in combination, either at a point or in a sequence. In such cases it can be helpful to resolve the forces, and their direction, into one unit.

Example 5.1
Two forces of known magnitude and direction act together in the same plane, as shown in fig. 5.2. Determine the combined force and its direction by graphical means.

A simple method, when considering two forces, is to construct both to scale so that they converge at a point. Using an adjustable square, a line parallel to each force is drawn to produce a parallelogram, know as a *force parallelogram*, the construction of which is shown in fig. 5.2. The diagonal from the point of convergence of the

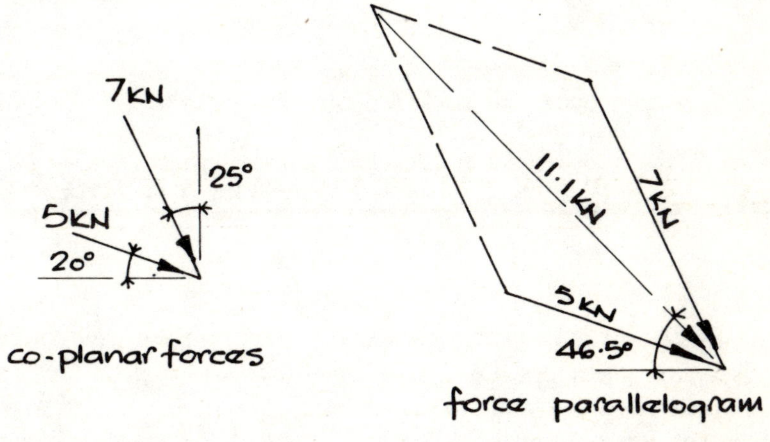

Fig 5.2 Resolution of co-planar forces by graphical method of force parallelogram

two forces is then produced. This diagonal gives both the magnitude and direction of the combined force.

This technique can be quite useful when a frame or component is subjected to forces acting in two directions. A typical situation can occur in a roof truss subjected to both wind loading, which causes a downward thrust on the windward side and an uplift on the leeward side of the truss, and dead load, which produces a downward force. Snow loading will also cause a downward force, and all three components can be seen in fig. 5.3.

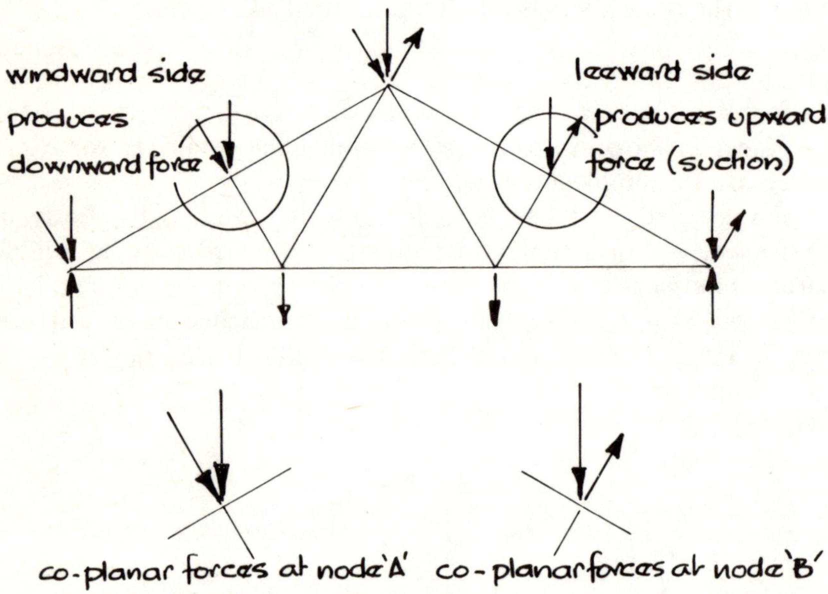

Fig 5.3 Effects of "wind loading" on roof truss

At this point it is worth pondering over a phrase used in each of the last two sentences, i.e. '. . . a dead *load* which produces a downward *force*' and 'snow *loading* will also cause a downward *force*'. At the beginning of chapter 4 we talked about *gravitational* force and the unit of force being the *newton (N)*. When we talk about *loads,* which are measured in kilograms, we are thinking about the mass of a material or how much it weighs rather than the force it exerts.

Thus a cubic metre of concrete which has a density of 2400 kg/m^3 would *weigh* 2400 kg. We do not often encounter a cubic metre of concrete in a structural component and we use density to arrive at its weight for a given thickness per square metre. So a floor slab of 125 mm thickness would weigh 0.125 x 2400 = 300 kg. To translate *weight* into *force*, we use a conversion factor based on gravitational acceleration. If we wished to be scientifically accurate, the figure used should be 9.80665 or 9.81, but for *structural* purposes a multiple of 10 can be used to convert kilograms to newtons. This is not because structural mechanics is slapdash, but we should always remember the materials that are used. No matter how accurate the production or assembly, our prime concern is that the materials exceed a *minimum* quality and that workmanship is within *limited* tolerances. Thus all work is likely to be slightly above these requirements. It should be noted also that our conversion factor is above actual value (i.e. 10 rather than 9.81, or 1.02 'on the safe side'). The gravitational force produced by the 125 mm slab would therefore be 300 x 10 = 3000 N/m^2. Rather than work in such large values we always use kilonewtons when dealing with *structural forces* so that the numbers are more manageable. Thus the slab would produce a force of 3 kN/m^2. What we have done is to multiply by 10 and then divide by 1000 so that we have actually divided the initial figure by 100. The conversion from mass to force therefore is *kilograms to kilonewtons (kg to kN) divided by 100.*

Now that we know how to determine gravitational forces, let us look back to our parallelogram of forces. Just as we are able to resolve two co-planar forces into one, we can also break down one inclined force into its vertical and horizontal components. This can also be very useful when *calculating* the effect of forces on a component or framework. We use exactly the same technique of the parallelogram but in reverse, i.e. we know the resultant and by producing vertical and horizontal lines to form an enclosing rectangle, we produce a parallelogram in which our *initial* inclined force forms the diagonal.

Example 5.2

Resolve the vertical and horizontal components of an inclined force as shown in fig. 5.4 by graphical means.

Figure 5.4 shows how simple this is and it also shows that we could, if required, arrive at a very accurate result by using trigonometry. Since we know the hypotenuse (our inclined force) of a

Fig 5.4 Triangle of forces graphically

right-angled triangle, we can quickly determine the other two sides using the sine and cosine of the angle (direction of force) thus:

sin 36° x 5 kN = vertical component = 2.94 kN
cos 36° x 5 kN = horizontal component = 4.05 kN

This is shown in fig. 5.5 as an alternative, using calculation. The figures obtained graphically of 3kN and 4 kN are reasonably accurate if they are to be used in further graphical analysis, but this example serves to show the importance of accurate draughting!

RESISTANCE TO FORCES

In talking about forces and resolving them, we must also consider how they might be resisted. When a force is *not* resisted, it will continue to move. This is best shown by considering what happens in space, where there is no gravity. If a spaceman goes outside the spacecraft without a lifeline or means of propulsion, such as a small booster jet, the force exerted in kicking against the craft will send him moving gently away from the space craft into space. To avoid this happening (we would lose spacemen that way), the lifeline

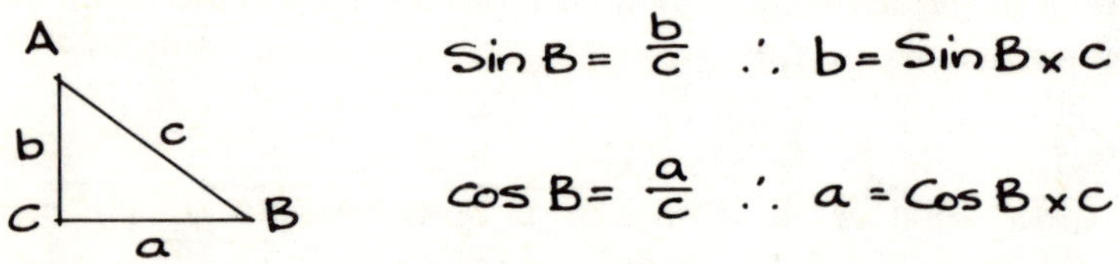

Fig 5.5 Triangle of forces by trigonometry

connects him to the craft so that he can pull his way back. This would not be a very satisfactory state of affairs in construction but happily the force of gravity produces a stabilising influence so that, sooner or later, any force is opposed by an equal and opposite force. For example, if a stone is dropped into a pond, it will eventually come to rest at the bottom of the pond since the bed will provide the necessary resistance.

In structural mechanics we are concerned with the 'sooner' rather than the 'later' resistance being developed, and if we know what resistance is required, in magnitude and direction, we can attempt to ensure that it is provided. Thinking back to our roof truss, we will try to ensure that the size of each component is sufficiently large to resist the imposed force and that the walls are sufficiently strong to support the roof. We will also attempt to avoid the roof blowing away in a gale.

If we consider a simple case of two co-planar forces of known size and direction, we can easily determine the resistance force, its size and direction. A graphical technique, using a triangle of forces, is shown in fig. 5.6. By drawing the forces in sequence, rather than converging, we are drawing two sides of a triangle. We then complete the triangle, continuing the direction around the triangle to its origin. This third side provides the magnitude and direction of the 'balancing' force to maintain equilibrium.

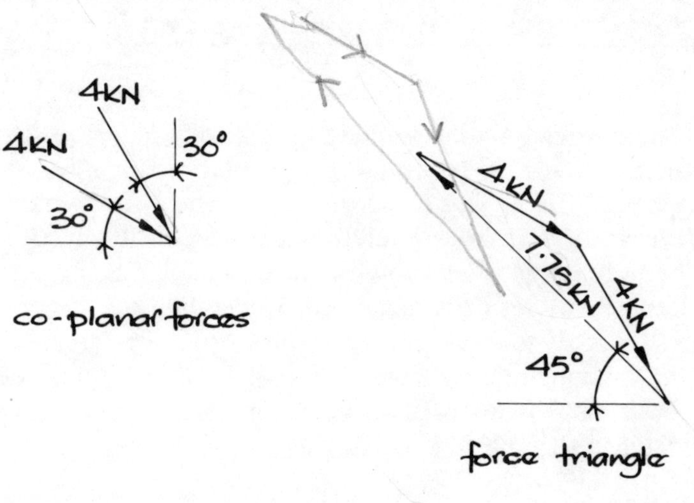

Fig 5.6 Triangle of forces to determine resistance force to co-planar forces

Example 5.3
Determine the magnitude and direction of force required to resist two co-planar forces of values as shown in example 5.1.

We construct the *triangle of forces*, as shown previously, and measure the magnitude and direction of the resistance force (see fig. 5.7). We will now see from the result that it is *exactly equal and opposite* to the resultant force that we determined when using the parallelogram of forces in example 5.1.

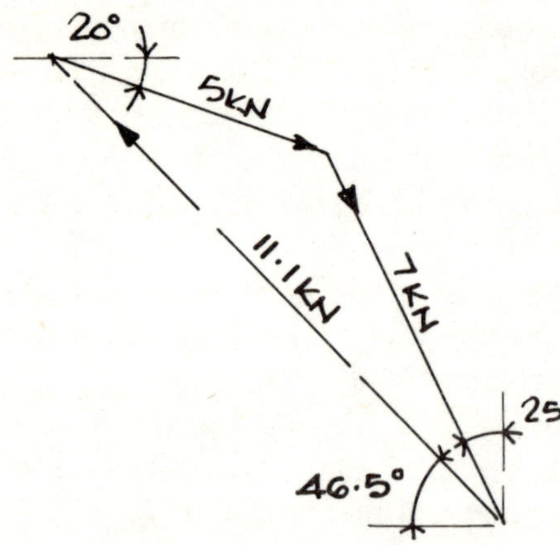

Fig 5.7 Alternative technique to that adopted in example 5.1 using triangle of forces

This technique can be developed further to incorporate three or more co-planar forces to determine the resistance force and thereby also the resultant force. This involves plotting the *force polygon*. The known forces are plotted round in sequence until all have been drawn. The polygon is then 'closed', the closing line being the resistance force, and the result is measured. Depending on the nature of the forces, the polygon may be of either 'open' or 'crossed' form (see fig. 5.8). Each in their own case will be equally correct. This technique can also be adapted to determine reaction forces, as will be shown later in the chapter.

STRESSES

If a structural component is located exactly along the line of force, the stress set up in the cross-section is termed a *direct stress*, which

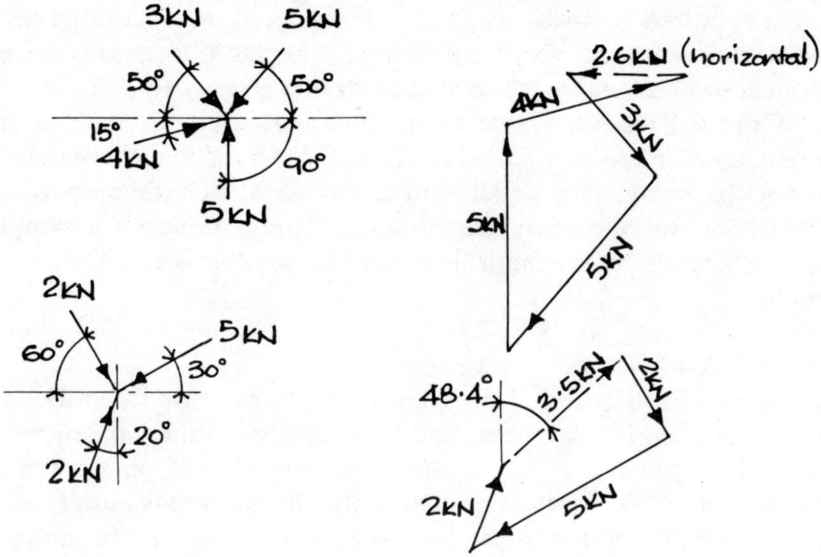

Fig 5.8 Examples of force polygons

will be either tensile or compressive. For example, a cable stay that supports sheet piling is under *tensile* stress whereas a strut supporting two opposite sides of a trench excavation is under *compressive* stress.

What then *is* direct stress? We know, from a logical commonsense reasoning, that a large structural unit is safer than a small one so cross-section must have something to contribute. We also know that some materials are stronger than others and can be of smaller cross-section. Quite simply, direct stress is the force divided by the cross-sectional area, i.e. the area of the cross-section at *right angles* to the line of force.

In the same way that applied forces equate to resistance forces, so applied stress must equate to the material stress. Since different materials possess different allowable working stresses, it is important to realise how these may be determined. Applied forces are not always constant: they may be present on some occasions but reduced or non-existent on others. For example, wind loads vary with barometric pressure; snow only falls in winter and may not even do so in some years. The material must be able to resist these forces when they occur, by absorbing them, and then relax to its former state when they are removed. It must be *elastic*.

If a compressive stress is applied, the material must have the ability to shorten without major distortion, whereas if a tensile stress is applied, the material must be able to stretch slightly to accommodate it. If the materials are able to do this and return to their original length upon removal of the stress, they have sufficient elasticity to accept the *strain*. It is possible to determine the elastic properties of a material by laboratory experiments. This is done on a sampling basis where a sample length is selected at random and tested to destruction.

Example 5.4
A mild steel rod is tested on a tensile rig to measure its elasticity. A force is applied in increments and the extension of the rod is measured for each increment of force. The results are plotted on a graph and the increments of force continued until the rod breaks under excessive strain. The points plotted are then joined together to produce a curve, which should follow a standard pattern. The force–extension graph shown in fig. 5.9 is typical of the material and has certain points along the curve which are of interest in telling us the behaviour of the material. Point *A* is the position at which the curve no longer continues in a straight line. This is known as the *limit of proportionality.* In other words, once this point has been passed, the extension is no longer proportional to the applied force. Point *B* shows a slight 'hiccup' in the behaviour of the material, which continues to extend slightly without additional force being applied. However, it then

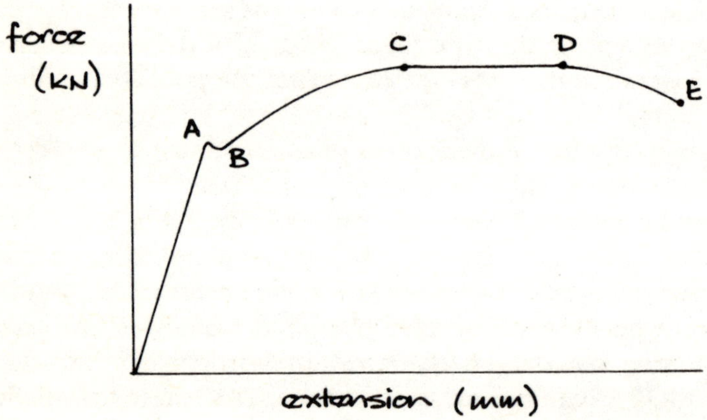

Fig 5.9 Force/extension graph

requires another increment to start the extension moving again. Point *C* shows another change in the behaviour of the rod, which continues to extend without any further force being applied. This is known as its *yield point*. Another change of curve is seen at point *D*, where the extension starts to accelerate, and even though the applied force is reduced, the extension will continue until the rod snaps. The ultimate failure point is seen at point *E*.

Points *A* and *C* are the important stages from a structural consideration and indicate valuable material properties. Alternative titles are sometimes given to these stages in the test, these being point *A* – *the elastic limit*, and point *C* – *the plastic limit*. If a material is to be subject to forces of considerable fluctuation, we tend to base its safe working stress on its elastic limit. If, however, the applied forces are reasonably consistent, we may base our calculations on its plastic limit. Alternatively, the peculiarities of the material may suggest which limit should be considered. For example, timber has a fair elastic range but a poor plastic range of behaviour so we use the elastic limit, whereas concrete has a poor elastic range but good plastic range of behaviour so we use the plastic limit. Whichever of the two bases we work from, the safe working stress adopted is taken at a point well below the limit achieved.

At this stage let us consider the behaviour of a material up to its elastic limit (or limit of proportionality). Up to this limit, the material behaves consistently in its relationship between force and extension, producing a straight-line graph, i.e. force/extension = elasticity. This, of course, applies to the specimen under test, which has a 'known' cross-section and a 'known' original length. We really need to apply the equation to the material for any cross-section and any length. To do this we translate the force into stress, thus: stress = applied force/cross-sectional area; and the extension into strain, that is: strain = extension (or contraction)/original length. This gives us the equation:

$$\frac{\text{Stress}}{\text{Strain}} = \text{Elasticity } (E)$$

The behaviour of a material within the limits was studied by many scientists and engineers, the more prominent among them being Hooke and Young. Because of this, the equation is often known as *Hooke's law* when in its original form of force/extension = elasticity, and the constant (*E*) is often described as *Young's modulus of elasticity*. Regardless of how we describe it, the important thing is that we remember the equation and the ways in which it can be transposed for particular applications. For example:

(a) In its original form it is useful as a sampling process for assessing the quality of a material.
(b) Strain = stress/elasticity is useful for assessing how much strain (or extension) a material can safely accept. (See example 4.9.)
(c) Stress = strain x elasticity is useful for determining, by non-destructive testing, the stress in a material subjected to an applied force. This can be achieved by measuring the strain with either a mechanical or electronic *strain gauge.*

Example 5.5
A ground anchor cable is subjected to a force of 100 kN. It has a cross-sectional area of 300 mm^2 and its modulus of elasticity is 230 kN/mm^2. Determine the extension for a length of 5 m.

We wish to find the extension so we will need to establish the *strain*. We know the elasticity but we need to calculate the stress. [*Note:* It is important to be consistent with the units in which we work. Fortunately, in this instance, our values are already in terms of kilonewtons and square millimetres, but be careful since this will not always necessarily be the case.]

$$\text{Stress} = \frac{\text{Applied force}}{\text{Cross-sectional area}} = \frac{100 \text{ kN}}{300 \text{ mm}^2} = \frac{1}{3} \text{ kN/mm}^2$$

Therefore:

$$\text{Strain} = \frac{\text{Stress}}{\text{Elasticity}} = \frac{1}{3 \times 230} = 1.45 \times 10^{-3}$$

We know that strain = extension/original length, so extension = strain x original length, which equals $1.45 \times 10^{-3} \times 5 \times 10^3 = 7.25$ mm.

Example 5.6
A timber strut, in a stud wall system, is checked for strain using a mechanical strain gauge. A reading of 7.5×10^{-4} is recorded. Given that E for timber = 8 kN/mm^2, determine:

(a) The stress;
(b) the force being exerted (100 x 50 stud).

In determining the stress we use the equation stress = strain x elasticity = $7.5 \times 10^{-4} \times 8$ kN/mm^2 = 6×10^{-3} kN/mm^2 or 6 N/mm^2.

Since we now wish to assess the force, we transpose the equation, stress = applied force/cross-sectional area, to read: stress x cross-sectional area = force, thus: 6×10^{-3} kN/mm^2 x 100 x 50 = 30 kN.

We have seen the relationship between force and stress and a similar relationship exists between force and pressure. When pressure is exerted on an object, its total effect can be translated into force, i.e. pressure x area = force. *Pressure* is not always consistent over the entire area of a structure, e.g. wind pressure is greater towards the top of a building, because of wind speed, and water pressure is greater at the bottom of a dam than at the top, because its pressure increases with depth. This is why divers need to adjust to change, in a decompression chamber, after working at depth (say, working on the substructure to an oil platform).

Because water is a liquid it will act equally, in all directions, on the face of its container. It is also dependent upon its 'head'. In Volume 2 we will consider water in motion but at this stage we need concern ourselves only with the 'static head' of water, which produces pressure. This can be used as a constant where the pressure brought about by 1 m head of water is 9.81 kN/m^2, or, in our case, 10 kN/m^2 for structural calculations.

Alternatively, we can work from first principles, which can be applied for any liquid, i.e. pressure = depth x density x gravitational factor. For water this can be seen as: depth x 1000 kg/m^3 x 9.81 (10), which, for 1 m = 1000 kg x 9.81 (10) $\div$ 10^3 $\triangleq$ 10 kN/m^2. This can be simplified so that knowing density in kilograms per cubic metre we divide by 100 (see the beginning of the chapter) and multiply by 1 m to arrive at pressure, i.e. oil pressure (density 750 kg/m^3) = 750 $\div$ 100 = 7.5 kN/m^2 for 1 m head.

Apart from being useful when determining the pressure on a container such as an oil storage tank, a cold water storage tank or a swimming pool, it is also helpful to establish the head required for a hot and cold water system to operate at an acceptable water pressure.

Ground-water will behave in the same way, developing pressure with the depth of water table, but can cause additional pressure if the ground becomes so saturated that the soil has no frictional resistance. The soil will then be in suspension so that the overall mixture will act as a liquid. This is why it is sensible to relieve the water pressure if a retaining wall does not have to be watertight.

Weephole details can be seen in chapter 7 of *Building and Civil Engineering Construction* Volume 1.

The development of earth pressure is affected by the angle of internal friction and this will be covered fully in Volume 2 of *Construction Science*. The force exerted, however, acts through its *centre of pressure*, which we can now study. This is quite often thought of incorrectly as *centre of gravity*, a term that is also incorrectly used when talking about *centres of area*.

A centre of area is a geometric property of a figure and for rectangles, triangles and trapeziums can be determined quite easily by graphics (see fig. 5.10). For a rectangle the intersection point of the diagonals shows that it occurs, quite logically, at the centre point from each corner. For the triangle it always occurs at the 'third' point measured at right angles to the two shorter sides.

A centre of gravity is really the centre of force of a series of downward or gravitational forces and is best determined by using *moments*. A moment is developed by a *couple*, which is a force x

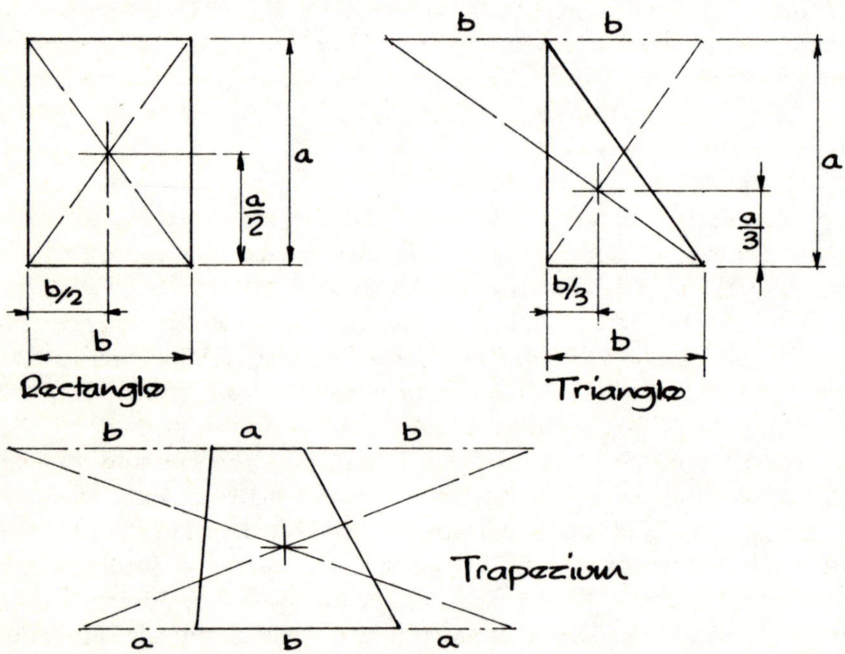

Fig 5.10 Graphical determination of centres of area

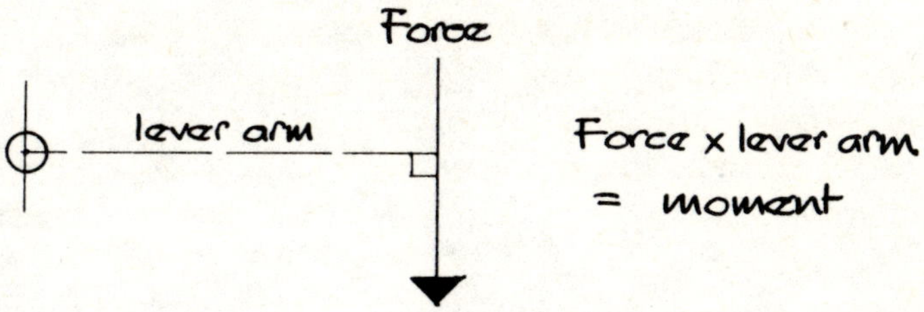

Fig 5.11 Creation of "Moment"

its *normal* (right-angled) distance from a given point (see fig. 5.11). This is usually calculated in kilonewton metres, these units being easy to handle when applied to structural problems. In the early part of this chapter we established that a force, or collection of forces, has an equal and opposite resistance. In the same way, a moment, or collection of moments, has an equal and opposite moment to retain stability. If this were not the case, objects would spin round in perpetual motion.

It is not important, when calculating the centre of gravity for a load system, to work from any predetermined point. So long as *all* moment calculations are taken around the *same* point, the answer, in relation to the load system, will be the same.

Example 5.7
A series of vertical forces, as shown in fig. 5.12, act together. Determine their centre of gravity.

To prove the statement that the point selected is irrelevant to the actual location of the centre of gravity (c. of g.), we will take two points, one outside the system and one inside, and by two sets of calculations will both prove the statement and check our results.

Answer 1: Select a point 2 m to the left of the 5 kN force, i.e. all downward forces are to the right of the point. In taking moments, all the downward forces will create clockwise (↷) moments whereas the resisting force (upward) will create an anticlockwise (↶) moment. Since all forces are in kilonewtons (kN) and all distances in metres (m), we do not need to suffix each value. We should, however, *always put the force first* since the moment value is expressed as kilonewtons times metres.

Thus the sum of the clockwise moments = (5 x 2) + (8 x 3.2) + (8 x 3.7) + (10 x 5) + (6 x 6.5) = 10 + 25.6 + 29.6 + 50 + 39 = 154.2 kNm↷.

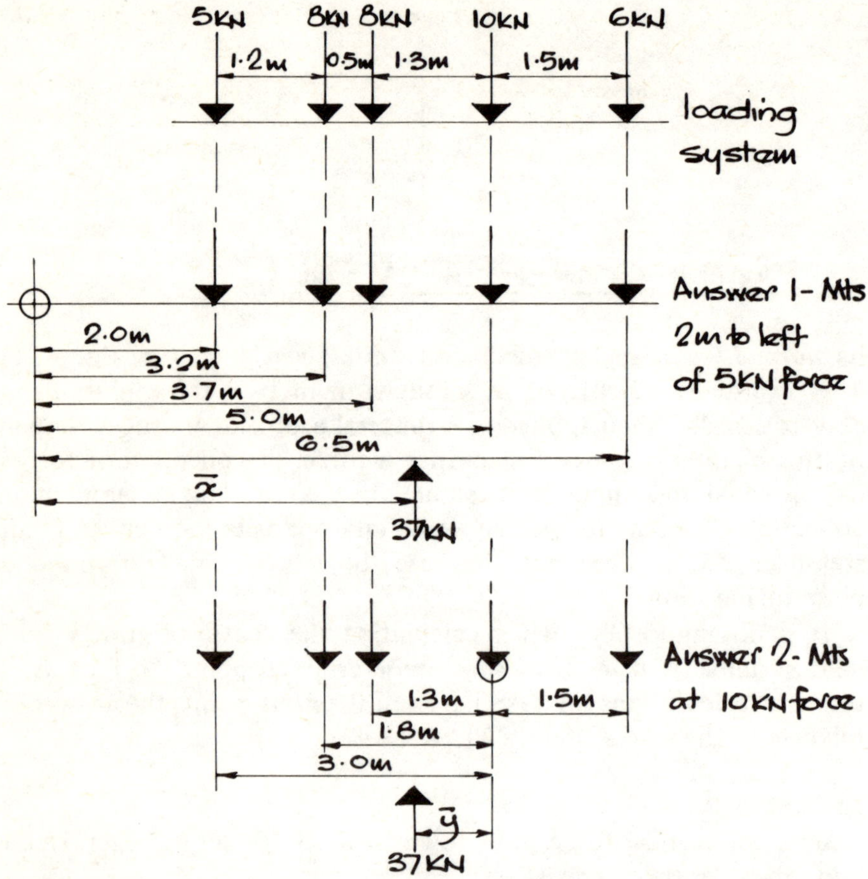

Fig 5.12 Determination of centre of gravity of a group of forces by taking moments

As stated before, the resistance force must equal the applied forces so 5 + 8 + 8 + 10 + 6 = 37 kN ↓ opposed by 37 kN↑. Since the resistance moment ↶ must also equal the applied moment, it will occur at a distance $\bar{x}$ to the right of the point chosen so that:

$$37 \times \bar{x} = \curvearrowleft \text{ kNm}$$

Therefore:

$$37\bar{x} = 154.2$$

Thus:

$$\bar{x} = \frac{154.2}{37} = 4.168 \text{ m}$$

This means that the centre of gravity is 2.168 m to the *right* of the 5 kN force, or 0.832 m to the *left* of the 10 kN force.

Answer 2: Take moments about the 10 kN force. This will mean that three loads act anticlockwise and one load acts clockwise, but the 10 kN force has *no* distance and therefore *no* moment.

Summation of applied moments = (5 x 3) + (8 x 1.8) + (8 x 1.3) $\curvearrowleft$ and (6 x 1.5) $\curvearrowright$ = 15 + 14.4 + 10.4 $\curvearrowleft$ + 9 $\curvearrowright$ = 39.8 $\curvearrowleft$ and 9 $\curvearrowright$ produces 30.8 $\curvearrowleft$. Resistance moment must be clockwise and will therefore mean that the resistance force is to the left of the 10 kN load.

Resistance moment = 37 x $\bar{y}$ $\curvearrowright$

so that:

$$37\,\bar{y} = 30.8$$

thus:

$$\bar{y} = \frac{30.8}{37} = 0.832 \text{ m}$$

The centre of gravity is 0.832 m to the left of the 10 kN force.

An important feature of the example just completed is that we have established some principles that are central to the bulk of structural mechanics theory. Since the summation of applied forces is equal to the resistance force (or forces), because we can resolve forces into their vertical and horizontal components, and clockwise moments are resisted by anticlockwise moments, we have established the basic stability equations that are used in all 'statically determinate' structural theory. Using the sign Σ for 'the summation of', we have the following set of equations:

$$\Sigma\downarrow = \Sigma\uparrow$$
$$\Sigma\rightarrow = \Sigma\leftarrow$$
$$\Sigma\curvearrowright = \Sigma\curvearrowleft$$

Once you have remembered these three equations you should have no great difficulty in understanding structural mechanics.

BEAM REACTIONS

Following example 5.3 we talked about force polygons and showed examples of these in fig. 5.8. We talked then about using this technique to determine beam reactions graphically. At that time we did not discuss beam mechanisms, but having now dealt with stability equations, let us demonstrate the use of both techniques. First, however, what is a *beam mechanism*? We should all know that horizontal structural components that support vertical forces are either beams or slabs and that slabs can be considered as a series of adjoining beams.

We have see that $\Sigma\downarrow = \Sigma\uparrow$ and so it is logical to reason that if we have a *simply supported beam*, carrying a series of loads producing downward forces at points along its span, we should know that the total of the end reactions is equal to the total of the downward forces. Let us first look at an example using graphical techniques to determine the reactions and then, using the stability equations, calculate the reactions by taking moments.

Example 5.8

A beam, simply supported at its ends, carries a series of *point loads* as shown at the top left-hand side of fig. 5.13. Determine the reaction forces R_1 and R_2

(a) by graphical means;
(b) by calculation.

Graphical solution

(a) Annotate the loading diagram by a letter between each force as shown.

(b) Plot a force line to a suitable scale to represent the downward forces, i.e. a to b = 20 kN; b to c = 30 kN; c to d = 40 kN. This line represents $\Sigma\downarrow$.

(c) Select a 'polar point' (o) to the right of the force line and connect each ordinate by an inclined force line. [*Note:* Graphical accuracy is always a problem and to avoid giving yourself too much work, use 30° for *ao* and *do* to establish the position o.]

(d) We know already that forces act in straight lines so produce the lines for R_1, the applied loads, and R_2, below the loading diagram.

(e) Construct a 'force polygon' starting with a line parallel to *ao* to cut the 20 kN force line. Now construct a line parallel to *bo* to cut the 30 kN force line. Next construct a line parallel to *co* to cut the 40 kN force line and finally a line parallel to *do* to cut the R_2 line.

(f) Close the polygon with a dotted line connecting R_2 to R_1 and translate this line parallel through the polar point to cut the force line as shown. This gives us point e on the force line so that *de* is

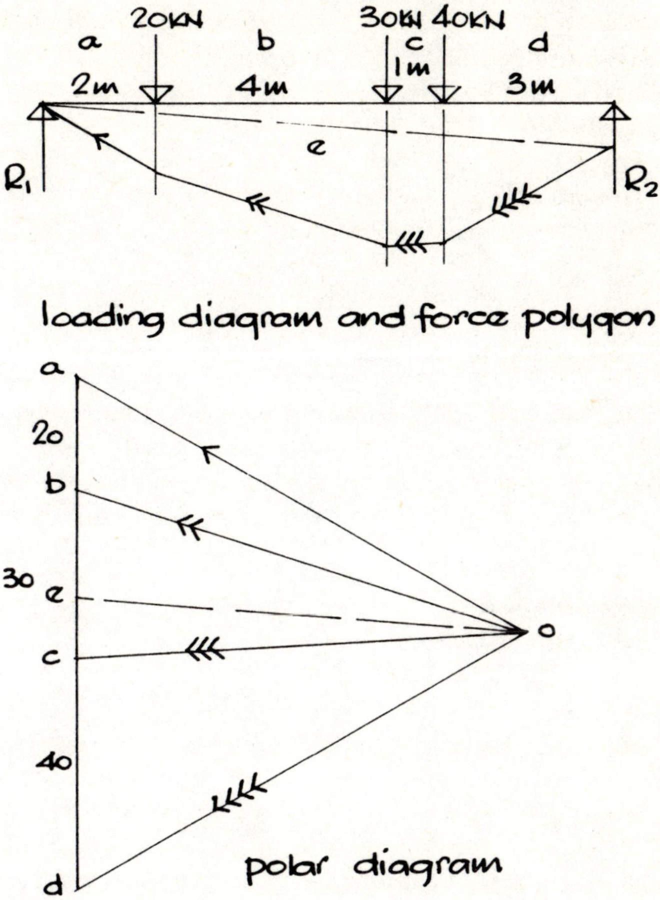

Fig 5.13 Graphical determination of beam reactions using polygon of forces

the value of R_2 and *ea* is the value of R_1. Scaling the results: R_1 = 40 kN; R_2 = 50 kN.

Calculation solution

(a) Take moments about one reaction to find the other reaction using $\Sigma \curvearrowright = \Sigma \curvearrowleft$.

$$\Sigma \curvearrowright = (20 \times 2) + (30 \times 6) + (40 \times 7)$$
$$= 40 + 180 + 280 = 500 \text{ kNm}$$
$$\Sigma \curvearrowleft = R_2 \times 10 \text{ kNm}$$
$$\Sigma \curvearrowright = \Sigma \curvearrowleft$$

therefore:

$500 = 10\,R_2$

$R_2 = 500 \div 10 = 50$ kN

(b) Summate forces using $\Sigma\downarrow = \Sigma\uparrow$

$\Sigma\downarrow = 20 + 30 + 40 = 90$ kN

$\Sigma\uparrow = R_1 + 50$ kN

$\Sigma\downarrow = \Sigma\uparrow$

therefore:

$90 = R_1 + 50$

$R_1 = 90 - 50 = 40$ kN

As you can see, both techniques produce the same results, if your draughtmanship is accurate, but you are encouraged to use the calculation method because you can always check your answer by taking moments about the other reaction, e.g. taking moments about R_2 to find R_1:

$\Sigma\curvearrowright = R_1 \times 10$ kNm

$\Sigma\curvearrowleft = (40 \times 3) + (30 \times 4) + (20 \times 8)$

$= 120 + 120 + 160 = 400$ kNm

$\Sigma\curvearrowright = \Sigma\curvearrowleft$

therefore:

$10\,R_1 = 400$

$R_1 = 400 \div 10 = 40$ kN

Unfortunately, not all beam mechanisms are simply supported at their ends. Also, not all forces are in the form of point-loaded systems. Figure 5.14 shows three types of end support, pinned, rollered and fixed; and three types of span, end supports, cantilevers and overhanging ends (a mixture of the first two). However, so long as a simple beam system is statically determinate, its reactions can be calculated by the technique used in example 5.8.

Cantilever beams, with only one reaction, must have a fixed support to stop them from 'hingeing' down. However, the reaction calculation is simple, because we only need to use $\Sigma\downarrow = \Sigma\uparrow$, the latter being the reaction. Beams with overhanging ends are also easy, as will be seen in the next example.

Example 5.9

A point-loaded system of beam mechanism with overhanging ends is shown in fig. 5.15. Determine the values of the reactions.

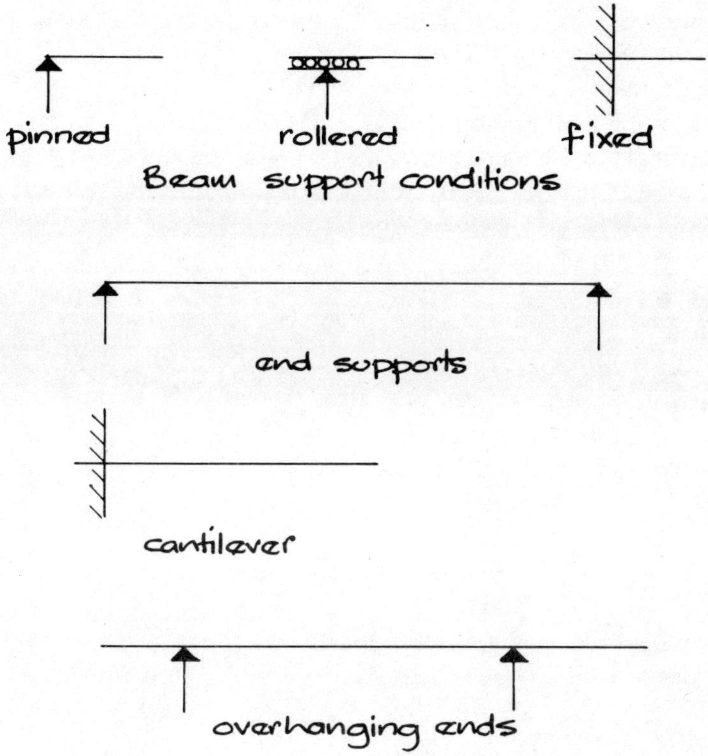

Fig 5.14 Statically determinate beam systems

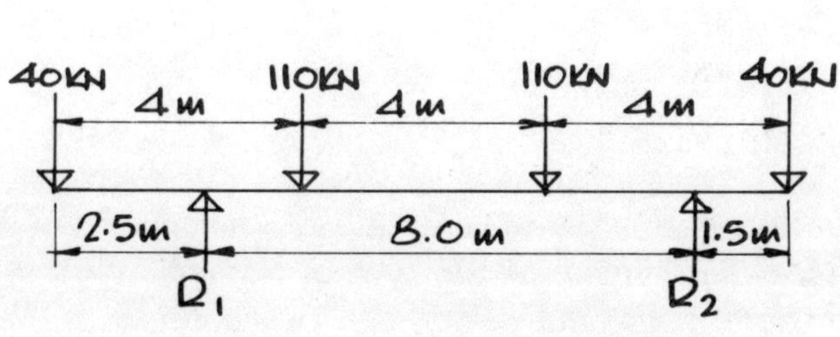

Fig 5.15 Beam system for example 5.9

Follow exactly the same calculation process used in the last example but, in addition, check your answer by taking moments about the second reaction.

$\Sigma\curvearrowright = \Sigma\curvearrowleft$ taking moments about R_1

$$\Sigma\curvearrowright = (110 \times 1.5) + (110 \times 5.5) + (40 \times 9.5)$$
$$= 165 + 605 + 380 = 1150 \text{ kNm}$$
$$\Sigma\curvearrowleft = (40 \times 2.5) + (R_2 \times 8)$$
$$= 100 + 8R_2 \text{ kNm}$$
$$\Sigma\curvearrowright = \Sigma\curvearrowleft$$

therefore:

$$1150 = 100 + 8R_2$$

so:

$$8R_2 = 1150 - 100 = 1050$$
$$R_2 = 1050 \div 8 = 131.25 \text{ kN}$$

$$\Sigma\downarrow = \Sigma\uparrow$$
$$\Sigma\downarrow = 40 + 110 + 110 + 40 = 300 \text{ kN}$$
$$\Sigma\uparrow = R_1 + 131.25 \text{ kN}$$
$$\Sigma\uparrow = \Sigma\downarrow$$

therefore:

$$300 = R_1 + 131.25$$
$$R_1 = 300 - 131.25 = 168.75 \text{ kN}$$

Check by taking moments about R_2.

$$\Sigma\curvearrowright = (R_1 \times 8) + (40 \times 1.5)$$
$$= 8R_1 + 60 \text{ kNm}$$
$$\Sigma\curvearrowleft = (40 \times 10.5) + (110 \times 6.5) + (110 \times 2.5)$$
$$= 420 + 715 + 275 = 1410 \text{ kNm}$$
$$\Sigma\curvearrowright = \Sigma\curvearrowleft$$

therefore:

$$8R_1 + 60 = 1410$$
$$8R_1 = 1410 - 60 = 1350$$
$$R_1 = 1350 \div 8 = 168.75 \text{ kN}$$

If you are wondering 'How can you get a beam system like this?', the fact is that it is typical of the type of framed structure comprising lock-up shops at ground floor, with two columns set back from the faces of the building supporting a beam which picks up columns to offices at first-floor level.

LOADING SYSTEMS

Immediately following the previous examples (5.8), we said that not all loading systems are of 'point-loaded' form. A very common form of loading is *uniform distributed loads (u.d.l.)*, which are produced by the dead load or self weight of the structure and, possibly, the imposed (or live) loads contained within it. 'Uniform' means that the rate is constant for a distance per metre run so that it is effectively of rectangular form. In determining the reactions we can plot the force as acting through its centre of gravity which we know, from earlier work in this chapter, is at the centre of the uniform distributed load.

Example 5.10
Two beam systems with uniform distributed and point loads are shown in fig. 5.16. Determine the reactions for each beam.

For beam (a) we can resolve the u.d.l. as 10 x 6 = 60 kN acting 3 m from R_2 and therefore 5 m from R_1.

$\Sigma \circlearrowright = \Sigma \circlearrowleft$ taking moments about R_1
$\Sigma \circlearrowright = 60 \times 5 = 300$ kNm
$\Sigma \circlearrowleft = R_2 \times 8 = 8R_2$ kNm
$\Sigma \circlearrowright = \Sigma \circlearrowleft$

therefore:

$300 = 8R_2$
$R_2 = 300 \div 8 = 37.5$ kN
$\Sigma\downarrow = \Sigma\uparrow$
$\Sigma\downarrow = 60$ kN
$\Sigma\uparrow = R_1 + 37.5$ kN
$\Sigma\downarrow = \Sigma\uparrow$

Therefore:

$60 = R_1 + 37.5$
$R_1 = 60 - 37.5 = 22.5$ kN

For beam (b) we can resolve the u.d.l. as 12 x 8 = 96 kN acting 2 m from R_1 and 6 m from R_2.

$\Sigma \circlearrowright = \Sigma \circlearrowleft$ taking moments about R_1
$\Sigma \circlearrowright = (96 \times 2) + (60 \times 10)$
$= 192 + 600 = 792$ kNm
$\Sigma \circlearrowleft = R_2 \times 8 = 8R_2$ kNm
$\Sigma \circlearrowright = \Sigma \circlearrowleft$

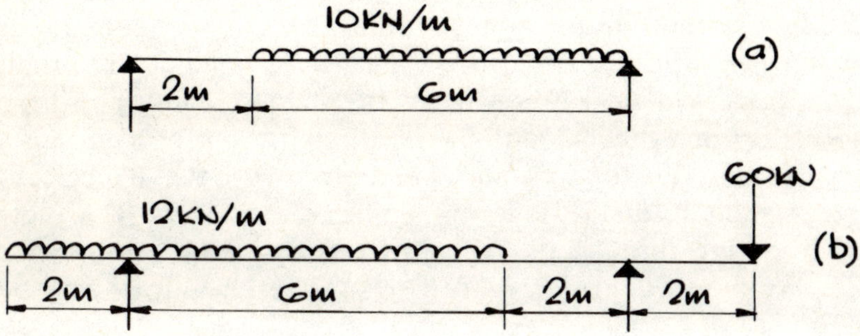

Fig 5.16 Beam systems for example 5.10

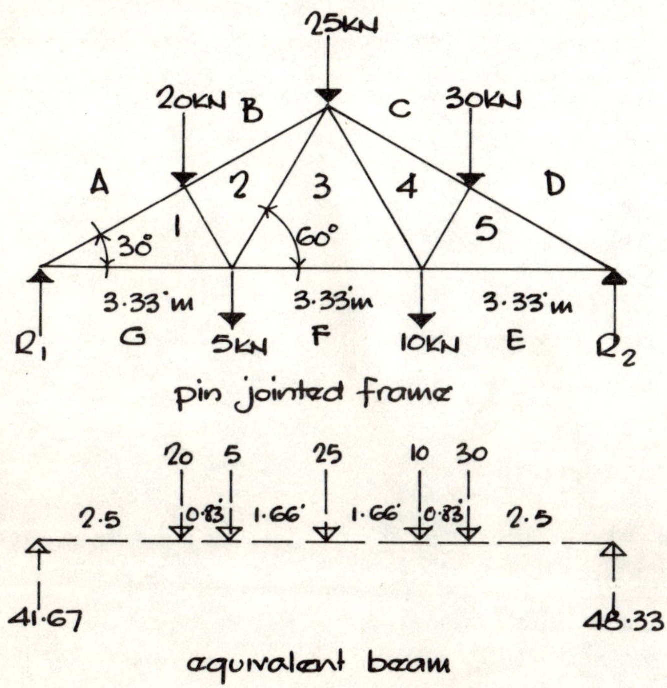

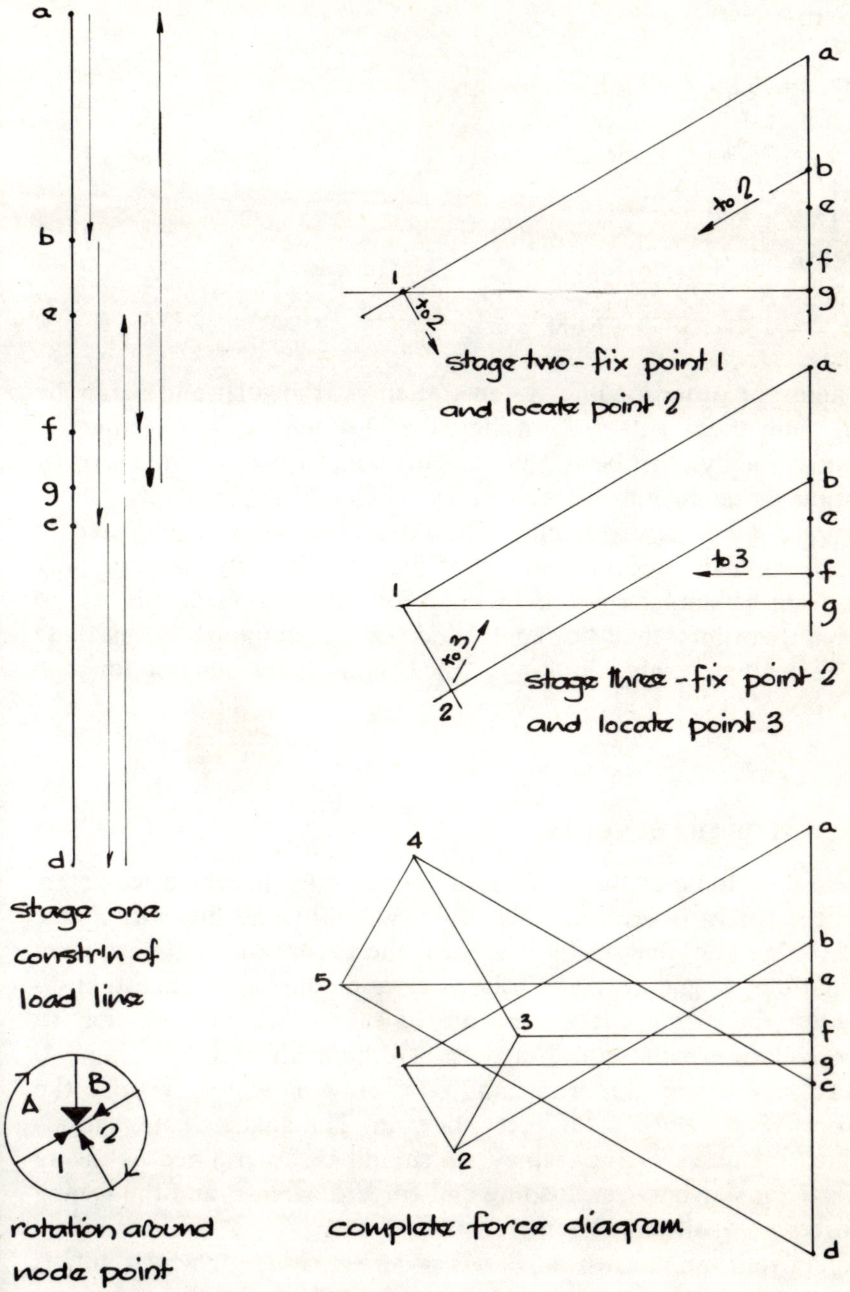

Fig 5.17 Graphic analysis of pin jointed framework illustrating development stages inc. nature of forces

therefore:

$792 = 8R_2$

$R_2 = 792 \div 8 = 99$ kN

$\Sigma\downarrow = \Sigma\uparrow$

$\Sigma\downarrow = 96 + 60 = 156$ kN

$\Sigma\uparrow = R_1 + 99$ kN

$\Sigma\downarrow = \Sigma\uparrow$

therefore:

$156 = R_1 + 99$

$R_1 = 156 - 99 = 57$ kN

There are other loading systems, such as triangular and parabolic loads, but these will be considered in Volume 2. The techniques illustrated above can be applied equally well to the determination of reaction forces to pin-jointed frameworks since, as previously stated, *all forces act in straight lines*. Thus if a framework is subjected to vertical applied forces, it will be calculated in exactly the same way as a beam of similar span. If inclined forces are applied, it is best to resolve them into their horizontal and vertical components and these will be dealt with when looking at the overall 'force analysis' for such conditions.

ANALYSIS OF FRAMEWORKS

Let us now look at the analysis of frameworks, in which we determine the forces in each member, since we will be dealing with *direct force* only. The analysis shown, with the exception of the reaction calculations, is graphical and follows certain conventions in establishing both the magnitude and nature of each force. It is possible to use calculation techniques but these will be dealt with in Volume 2.

Accuracy is very important and poor drawing will show up in the solution when 'closing' the force diagram. The angles of the various inclined members of the framework should be ascertained to ensure correct transfer between loading and force diagrams, and the scaling of forces is equally important.

A convention, known as *Bow's notation*, is used, which entails lettering the space between each force, in a clockwise direction, and numbering the space between each member of the framework from left to right. This will be shown in the examples and the reasons will become apparent as the work proceeds.

Example 5.11
Figure 5.17 shows a pin-jointed framework, with two simple supports, subject to a series of point loads at joint intersections (*node points*) and the graphical method by which the forces in the members are resolved. The reactions are calculated, rather than solved graphically, though you may practise the graphical technique if you wish. The reason that the loading is not symmetrical is because the right-hand side is fitted with 'solar collector panels' and their associated pipework. [*Note:* If the loading had been symmetrical, it would have been unnecessary to use moments to calculate the reactions as they would each have been equal to half the applied forces.]

The dotted-line diagram shows the equivalent beam from which we can determine the reactions. The dimensions are arrived at by knowing the properties of the 60°, 30°, 90° triangle (more of which is explained in *Construction Mathematics* Volume 1) and have been projected for each of the point load forces. Reaction calculations are:

$\Sigma \circlearrowright = \Sigma \circlearrowleft$ taking moments about R_1

$$\Sigma \circlearrowright = (20 \times 2.5) + (5 \times 3.33) + (25 \times 5) + (10 \times 6.67) + (30 \times 7.5)$$
$$= 50 + 16.65 + 125 + 66.67 + 225$$
$$= 483.32 \text{ kNm}$$
$$\Sigma \circlearrowleft = R_2 \times 10 = 10R_2 \text{ kNm}$$
$$\Sigma \circlearrowright = \Sigma \circlearrowleft$$

therefore:

$$483.32 = 10R_2$$
$$R_2 = 483.32 \div 10 = 48.33 \text{ kN}$$
$$\Sigma\downarrow = \Sigma\uparrow$$
$$\Sigma\downarrow = 20 + 5 + 25 + 10 + 30 = 90 \text{ kN}$$
$$\Sigma\uparrow = R_1 + 48.33 \text{ kN}$$
$$\Sigma\downarrow = \Sigma\uparrow$$

therefore:

$$90 = R_1 + 48.33$$
$$R_1 = 90 - 48.33 = 41.67 \text{ kN}$$

With these forces known, the force line can be plotted fully starting $ab = 20\downarrow$, etc. so *ga* should read as 41.67 kN.

To construct the force diagram, a line parallel to member A (on the loading diagram) is drawn through point *a* (on the force line) and a line parallel to G is drawn through point *g*. Where they cross establishes the position of point 1 on the force diagram. Next, a line parallel to B–2 is drawn through *b* to meet a line parallel to 1–2, drawn through 1. This establishes point 2. The construction is continued, drawing a line parallel to F–3 through *f*, and one parallel to 2–3 through 2, to establish point 3. A line parallel to C–4 through *c* and 3–4 through 3 locates point 4, and a line parallel to E–5 through *e* and 4–5 through 4 locates point 4. If the drawing is accurate, a line parallel to D–5 through *d* should meet point 5.

[*Note:* For this first example each stage has been shown but for future examples and in practice, all linework would be developed on the one diagram as is shown in locating point 5.]

With the force diagram complete we can now scale off the forces in the members but first we will determine their nature. This will show the importance of using a convention such as Bow's notation. We concentrate our attention around a node point and starting at the 20 kN force we move *clockwise* around the joint and refer continuously to the force diagram. *The direction of the force line is represented by an arrow at the node point.*

We already know that A to B travels downwards on the force line; B to 2 travels downwards towards the left; 2 to 1 travels upwards towards the left and 1 to A travels upwards towards the right. Thus *all* arrows point *in* towards the joint.

Moving to the 5 kN force again we move *clockwise* around the joint. As we know, F to G travels downwards on the force line; G to 1 travels right to left; 1 to 2 travels downwards to the right; 2 to 3 travels upwards to the right and 3 to F travels left to right. Thus all arrows except 1–2 point away from the joint.

The process is repeated for each joint and for this example the build-up at each node point is shown. For further examples, and in practice, all work would be carried out on the loading diagram.

The effect of plotting these direction arrows is that they illustrate, quite clearly, the nature of the force in each member. Where an arrow *pushes into* the joint the member is in *compression*. Where an arrow *pulls away from* the joint the member is in *tension.*

You will notice that the arrows at each end of a member are pointing in opposite directions. This *must* happen because if one end showed compression and the other tension, the member would get shot out of the framework in the direction of its arrows.

Now that we know the nature of each force and can measure its magnitude from the force diagram, we record each value. We could note the information on the force diagram but it is more useful, and easier to refer to, if we tabulate (see table 5.1). It is usual to start with the external members and then to record the internal members.

Table 5.1 Force values in example 5.11

Member	Compression	Tension
A–1	83.3	–
B–2	73.3	–
C–4	81.7	–
D–5	96.7	–
G–1	–	72.2
F–3	–	52.2
E–5	–	83.7
1–2	17.3	–
4–5	26.0	–
2–3	–	21.6
3–4	–	34.6

From a practical point of view we also group the members so that we can see, quite clearly, which is the determining force for sizing up the members in the framework. This is important because, although the loading arrangement is not symmetrical, the shape of the frame is symmetrical. It could therefore be placed incorrectly, during construction, if members on one side were different to those on the other, with quite disastrous results. Also the 'rafter' section, A–1 and B–2, would be formed from one length of material rather than two. Similarly the 'tie' joist section, G–1, F–3 and E–5, would be one length of material.

Example 5.12

A cantilevered roof truss, as shown in fig. 5.18, has its top reaction tied horizontally into a brick wall. We will analyse the forces graphically, showing that the nature of the framework produces both vertical and horizontal reaction components even though the applied forces are vertical.

We have again used a 30° angled truss, not because this is the only angle used but to facilitate the exercise in understanding its graphic analysis and to make the figures simple for calculation of its reactions. (This illustrates a further benefit of knowing particular properties of the 60°, 30°, 90° triangle.)

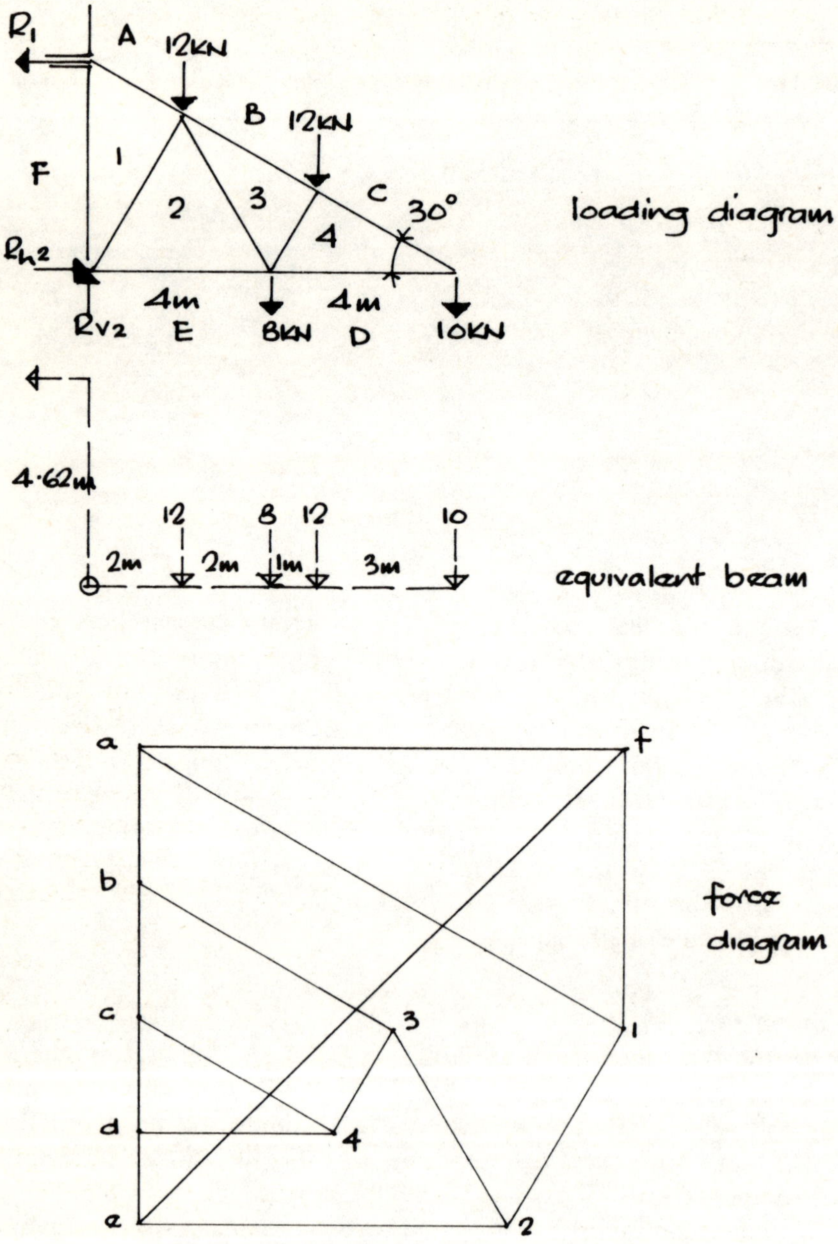

Fig 5.18 Cantilevered roof truss

$\Sigma \curvearrowright = \Sigma \curvearrowleft$ taking moments about R_2. [*Note:* We take moments about R_2 since we already know the direction of R_1.]

$\Sigma \curvearrowright = (12 \times 2) + (8 \times 4) + (12 \times 5) + (10 \times 8)$

$= 24 + 32 + 60 + 80 = 196$ kNm

$\Sigma \curvearrowleft = R_1 \times 4.62$ kNm

$\Sigma \curvearrowright = \Sigma \curvearrowleft$

therefore:

$196 = 4.62\,R_1$

$R_1 = 42.42$ kN

$\Sigma\downarrow = \Sigma\uparrow$

$\Sigma\downarrow = 12 + 8 + 12 + 10 = 42$ kN

$\Sigma\uparrow = R_{v2}$

therefore:

$R_{v2} = 42$ kN

$\Sigma\rightarrow = \Sigma\leftarrow$

$\Sigma\rightarrow = R_{h2}$

$\Sigma\leftarrow = R_1 = 42.42$ kN

therefore:

$R_{h2} = 42.42$ kN

Using trigonometry we can establish that the inclined reaction is 59.7 kN at 45.3° to the horizontal. Alternatively, using the force parallelogram, we may determine graphically a figure of about 60 kN at 45°. Either set of figures is acceptable, being within a 1% degree of error, if we remember that our 'gravity' figure of 10 is just below a 2% error. We now plot the force diagram as before, with the exception that our force 'line' is now a force triangle, and determine the nature and magnitude of the forces. The graphical work is developed in fig. 5.18 and the results are given in table 5.2.

You will notice that on this occasion the force diagram is very different. This is because of (a) the inclined reaction, and (b) the cantilever effect. Also, the rafter section is in tension whereas the horizontal section is in compression. This reversal is also caused by the cantilever.

Our final example of frameworks introduces the concepts of wind loads, roller support at one end and composite loads resolved graphically for analysis and into their vertical and horizontal components for reaction calculations.

Example 5.13

A northlight roof truss, of 7.2 m span, is subject to wind forces

Table 5.2 Force values in example 5.12

Member	Compression	Tension
A–1	–	49
B–3	–	26
C–4	–	20
D–4	17.3	–
E–2	32.3	–
F–1	24.5	–
1–2	20.2	–
3–4	10.4	–
2–3	–	19.6

creating pressure and suction as shown in fig. 5.19. It is also subjected to downward forces from dead and live loads. Reaction R_2 is a roller bearing to eliminate horizontal reaction forces. R_1 is composed of both vertical and horizontal force components. The calculations for R_2 can utilise the forces as shown acting at right angles through R_1. [*Note:* Where a force passes through the fulcrum (R_1), it has no distance (or lever arm) and therefore no moment.]

$\Sigma \curvearrowright = \Sigma \curvearrowleft$ taking moments about R_1

$$\Sigma \curvearrowright = (4 \times 3.6) + (10 \times 1.8) + (12 \times 4.5) + (6 \times 7.2)$$
$$= 14.4 + 18 + 54 + 43.2 = 129.6 \text{ kNm}$$
$$\Sigma \curvearrowleft = (4 \times 3.12) + (2 \times 6.24) + (R_2 \times 7.2)$$
$$= 12.48 + 12.48 + 7.2R_2 = 24.96 + 7.2R_2 \text{ kNm}$$
$$\Sigma \curvearrowright = \Sigma \curvearrowleft$$

therefore:

$$129.6 = 24.96 + 7.2R_2$$
$$7.2\,R_2 = 129.6 - 24.96 = 104.64 \text{ kNm}$$
$$R_2 = 104.64 \div 7.2 = 14.53 \text{ kN}$$

It is at this stage that the vertical and horizontal components are required and these are shown in fig. 5.20.

[*Note:* (a) Because the force at B–C comprises *three* components, *two* parallelograms of force are drawn, first the 10 kN and 4 kN, then the resultant of these with the 2 kN component. (b) The upward forces produce a differently shaped parallelogram since they move upwards into the diagram.

$\Sigma\rightarrow = \Sigma\leftarrow$
$\Sigma\rightarrow = 3.46 + 4.46 + 2 + 1 = 10.92$ kN
$\Sigma\leftarrow = R_{h1}$
$\Sigma\rightarrow = \Sigma\leftarrow$

therefore:

$10.92 = R_{h1}$
$\Sigma\downarrow = \Sigma\uparrow$
$\Sigma\downarrow = 6 + 10.27 + 8.54 + 4.27 = 29.08$
$\Sigma\uparrow = R_{v1} + 14.53$
$\Sigma\downarrow = \Sigma\uparrow$

therefore:

$29.08 = R_{v1} + 14.53$
$R_{v1} = 29.08 - 14.53 = 14.55$ kN

From fig. 5.20, $R_1 = 18.19$ at 36.9° to the vertical.

These values can now be translated on to the force triangle from which the force diagram is produced, as before, in fig. 5.21. The results are given in table 5.3.

Table 5.3 Force values in example 5.13

Member	Compression	Tension
B–1	10.2	–
C–2	14.5	–
D–3	20.5	–
F–1	–	12.4
F–3	–	18.8
1–2	–	6.4
2–3	6.4	–

[*Note:* In this example we included the forces at the junctions between rafter and bottom boom (A–B and D–E) because of their directions. In previous examples there was no requirement for this, since they were vertical forces being translated directly to the supports.]

This completes the work on graphical analysis and, as the examples showed (particularly 5.12 and 5.13), the accuracy of draughting and

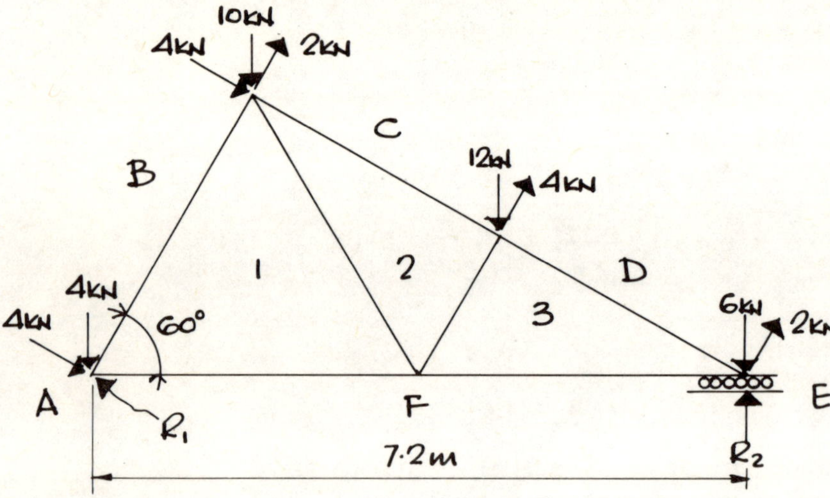

Fig 5.19 Loading diagram - North light truss

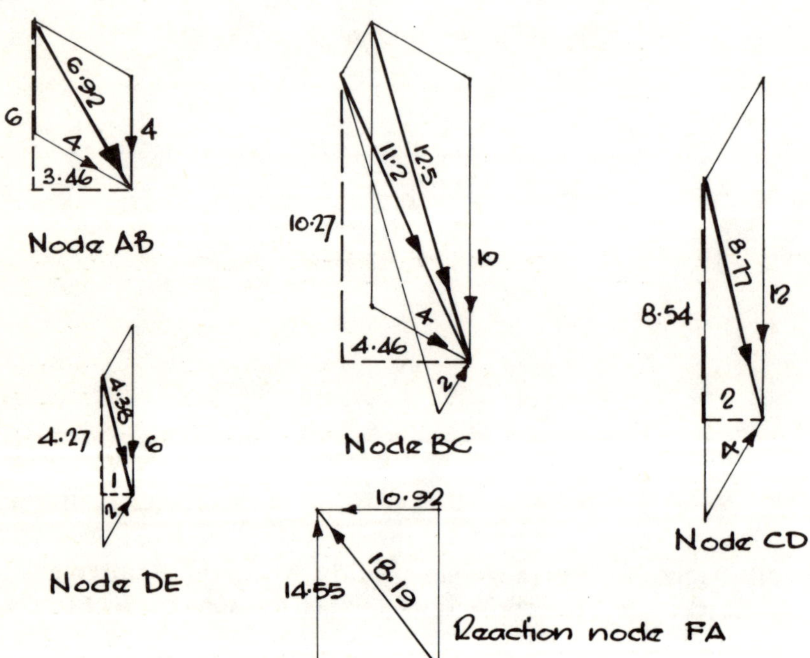

Fig 5.20 Resolution of force components

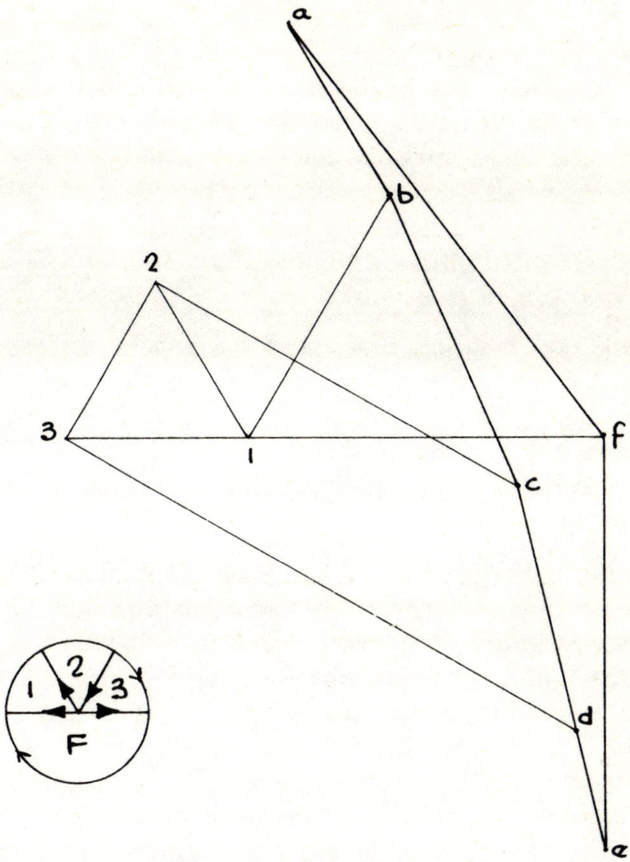

Fig 5.21 Force diagram - North light truss

scaling of forces obtained is of critical importance. You will notice that all three examples give tabulated values to *one decimal place only*. It is not necessary to be more exact than this in structural analysis for the reasons given earlier.

SHEARING FORCES AND BENDING MOMENTS

The pin-jointed frames were concerned only with direct force values, each member being in either compression or tension. In beams, which do not possess this open structure, the effect is to produce *shearing forces* and *bending moments* and these will be studied in this final part of the chapter.

It is possible to analyse beams graphically, as the reaction examples earlier in the chapter show, but it is much more practical to use calculations providing 'sketch' diagrams (not to scale) to relay the information visually. As stated earlier, and as the examples will show, this does not entail any form of advanced mathematics.

Example 5.14

A beam of single span, simply supported, is subjected to a series of point loads. Knowing its reactions:

(a) calculate the shearing forces at each change of force; and

(b) plot a sketch of the shearing force diagram to represent these values;

(c) calculate the bending moments at each change of force; and

(d) plot a sketch of the bending moment diagram to represent these values.

The loading diagram (fig. 5.22) is the same as that used earlier in example 5.8, which is why we already know the reactions.

We stated previously that every force is resisted by an equal opposing force. This is true not only of external forces but also of those internal forces that prevent a structural member from breaking. The resistance will be dealt with in Volume 2, but let us now see how the shearing forces can be established very simply. Again we use a *convention* in order to present information in a standard format. This reduces the risk of error in communicating such information.

To construct a shearing force diagram we *start from the left* and record each force in the direction it occurs and at the magnitude given. Where there is *no force* present there is *no change* on the diagram, i.e. the line plotted is parallel to the beam (horizontal). Thus the first movement from the baseline is 40 kN upwards (R_1). We then travel horizontally for 2 m until we meet the 20 kN force

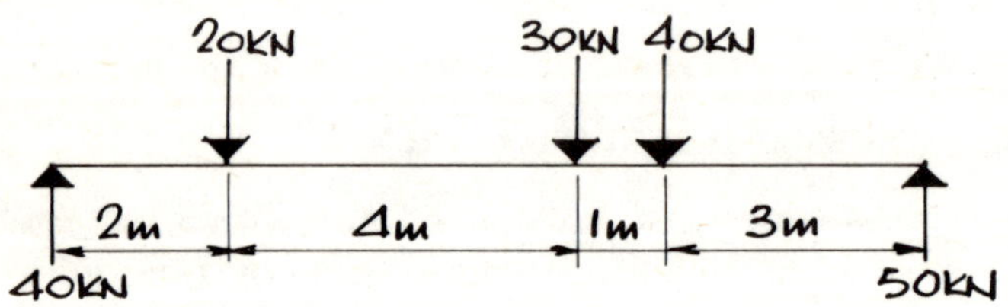

Fig 5.22 Loading diagram

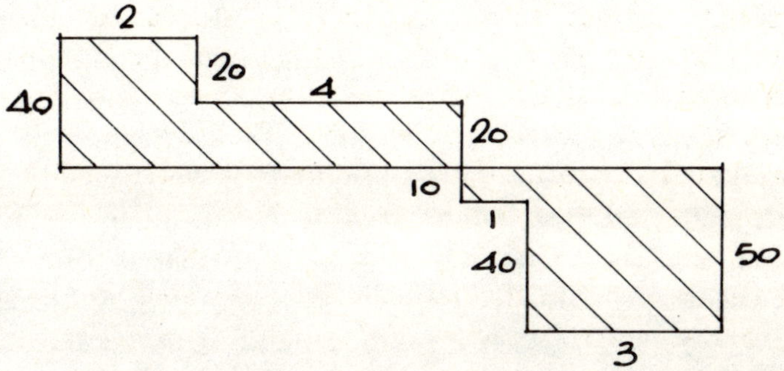

Fig 5.23 Shearing force diagram

and move downwards 20 kN. We travel another 4 m at the horizontal and then move down 30 kN. This brings us below the baseline. A further 1 m horizontal line brings us to the 40 kN force, which we plot downwards, followed by 3 m horizontally, at which point we reach the second reaction (R_2) and plot upwards 50 kN, which closes the diagram, bringing us back to the baseline. The results are shown on fig. 5.23. You will notice that the diagram is a series of rectangles. This is always the case for point-loaded systems since all the force lines are vertical.

We now come to determining the bending moment values and a second *convention*. Having travelled from left to right with the shearing force diagram, we now move down and return from *right to left*. In doing this we will produce a *tension force* diagram for the bending moment values.

There are two relationships between shearing force and bending moment values, the first being:

> 'The area of the shearing force diagram *to* any point equals the bending moment value *at* that point.'

Thus at the 40 kN force, the area, all below the line, is 50 x 3 = 150 kNm. We plot this below the baseline on the bending moment diagram (see fig. 5.24) measured from the baseline directly below the force.

Travelling along to the 30 kN force, the area is 150 kNm plus the smaller rectangle of 10 x 1 = 10 kNm giving a total of 160 kNm. Again we plot this below the baseline directly under the 30 kN force.

We now move to the 20 kN force. The third rectangle is above the baseline of the shearing force diagram so its area must be *subtracted* from the value plotted so far, i.e. 160 − (20 x 4), which equals 160 − 80 = 80 kNm, still below the baseline of the bending moment diagram. Finally, we arrive at the reaction, the last rectangle still being above the baseline of the shearing force diagram to give a running total of 80 − (40 x 2) = 80 − 80 = 0. We then join all the points of bending moment values recorded to produce the bending moment diagram (fig. 5.24), which is of triangulated form. This is always the case in point-loaded systems. We can now note the *second* relationship between shearing force and bending moment which is:

> 'The point (or points) of *zero* shear always coincide(s) with the point (or points) of *maximum* bending moment value.'

In this example, the baseline is crossed *once* (at the 30 kN force) so there is only *one* maximum bending moment value. On overhanging beams the baseline is crossed two or three times, which produces more maximum values for those stretches of beam. The three figures 5.22 (loading), 5.23 (shearing) and 5.24 (bending) are plotted below each other to show their interrelationship.

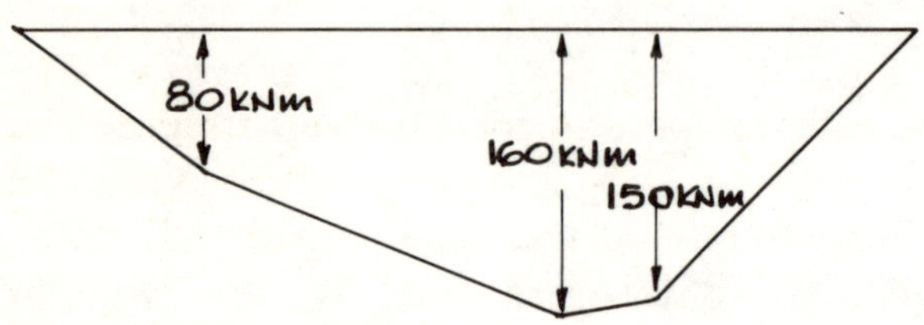

Fig 5.24 Bending moment diagram

We could have taken moments at each point of force to determine the bending moment values, had we wished, but this would not have told us whether the value should be plotted above or below the line. For example:

Moment at 20 kN force, reading left = 40 x 2 = 80 kNm ↻ or reading right = (50 x 80) − (30 x 4) − (40 x 5)
= 400 − 120 − 200 = 80 kNm ↺

Similarly, at the 30 kN force,
reading left B.M. = (40 x 6) − (20 x 4)
= 240 − 80 = 160 kNm ↻
or, reading right = (50 x 4) − (40 x 1)
= 200 − 40 = 160 kNm ↺

These can be useful as a check to your area summations.

In the next example we will look at the effect of uniformly distributed loading and overhanging ends to show that the principles remain constant for any beam system.

Example 5.15

A beam of 12 m length is simply supported at its left-hand end and at a point 9 m along its length. It sustains a uniformly distributed load, over its entire length, which exerts a force of 4 kN/m. Determine the maximum shearing force and bending moment values and sketch the diagrams to show where they occur.

We must retain a logical approach to the problem, establishing a procedure and checking our answers at each stage, i.e.

(a) Determine reactions by taking moments about one reaction and then summating forces.
(b) Check by taking moments about the other reaction.
(c) Plot the shearing force diagram from left to right. If it closes, the solution is self-checking.
(d) Summate the shearing force diagram to establish bending moment values.
(e) Check as required by taking moments.

We know from previous study that a u.d.l. force system is rectangular and acts through its centre. In this example the force is $4 \times 12 = 48$ kN and it acts 6 m from R_1 and 3 m from R_2.

$\Sigma \curvearrowright = \Sigma \curvearrowleft$ taking moments about R_1

$\Sigma \curvearrowright = 48 \times 6 = 288$ kNm

$\Sigma \curvearrowleft = R_2 \times 9 = 9R_2$

$\Sigma \curvearrowright = \Sigma \curvearrowleft$

therefore:

$288 = 9R_2$

$R_2 = 288 \div 9 = 32$ kN

$\Sigma\downarrow = \Sigma\uparrow$

therefore:

$48 = R_1 + 32$

$R_1 = 48 - 32 = 16$ kN

Check

$\Sigma \curvearrowright = \Sigma \curvearrowleft$ taking moments about R_2

$\Sigma \curvearrowright = R_1 \times 9 = 9R_1$

$\Sigma \curvearrowleft = 48 \times 3 = 144$ kNm

$\Sigma \curvearrowright = \Sigma \curvearrowleft \quad 9R_1 = 144$

$R_1 = 16$ kN

Plot the shearing force diagram.

[*Note:* Because the loading is of rectangular form, it falls at a constant rate, i.e. 4 kN/m. the diagram will therefore be triangulated (see fig. 5.25).]

Starting from the left, we move directly up 16 kN. We then fall at a constant rate until we reach the second support, i.e. $9 \times 4 = 36$ kN so that by this point we are $36 - 16 = 20$ kN *below* the baseline. We then move up directly 32 kN, which brings us 12 kN *above* the baseline. For the final 3 m we again fall at a constant rate $4 \times 3 = 12$ kN, which brings us back to zero at the end of the beam. The diagram closes, thus checking itself. On this occasion the baseline is crossed at two positions: R_2 and a point somewhere between R_1 and R_2. This is determined quite easily by dividing the distance fallen (16 kN) by the rate of fall (4 kN/m). Points of zero shear therefore occur at 4 m from R_1 and at R_2.

We now start to plot the bending moment values. At R_2 the triangle of shearing force is above the baseline so the area ($12 \times 3 \div 2$) is located above the baseline of the bending moment diagram.

B.M. value $= 36 \div 2 = 18$ kNm

The next point of zero shear occurs 5m further along the beam. The triangle is below the baseline and will be of opposite value to the first B.M., i.e. $(20 \times 5 \div 2) - 18 = 50 - 18 = 32$ kNm plotted below the line. Continuing to R_1, the final triangle is again above the line and must also be subtracted, i.e. B.M. $= 32 - (16 \times 4 \div 2) = 32 - 32 = 0$.

In checking by taking moments we must remember that each section of load acts through its centre of force. B.M. at 4 m from R_1, reading left $= (16 \times 4) - (16 \times 2) = 32$ kNm

B.M. at R_2, reading right $= 12 \times 1.5 = 18$ kNm

Thus we have two maximum bending moments:

32 kNm over the span (tension in the bottom)

18 kNm over the internal support (tension in the top)

The shape of the bending moment diagram is also different to that produced by point loads. When we summate the area of a triangle (in small increments), the values that we obtain produce a *parabolic* curve. All u.d.l. systems produce this type of curve. It is not critical that the diagram be drawn exactly to shape or scale, since most of the information we might require has already been determined. Some additional development work will, however, be covered in volume 2.

For this example all three diagrams, loading, shearing, and bending,

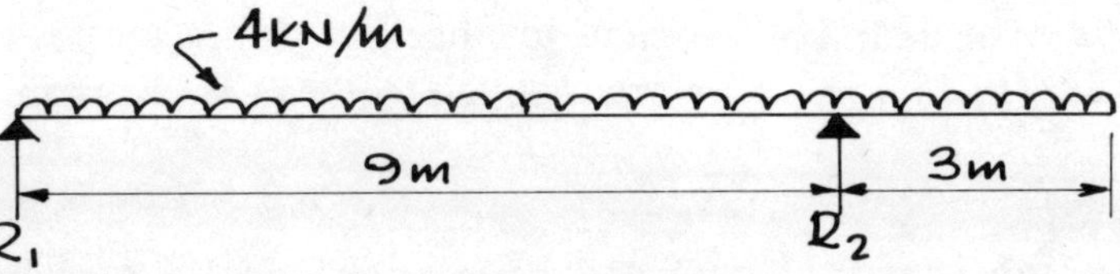

Loading diagram

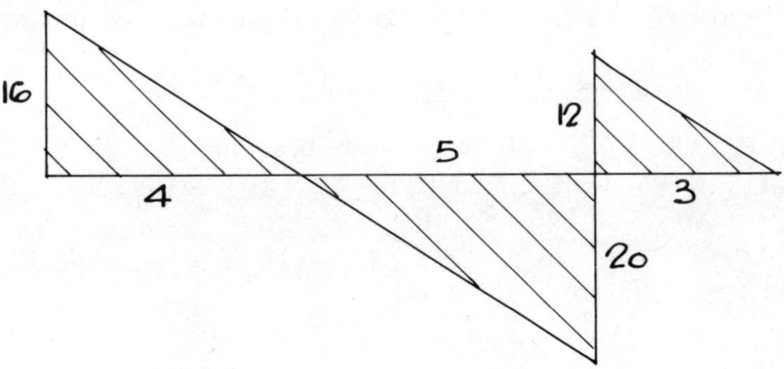

Shearing force diagram

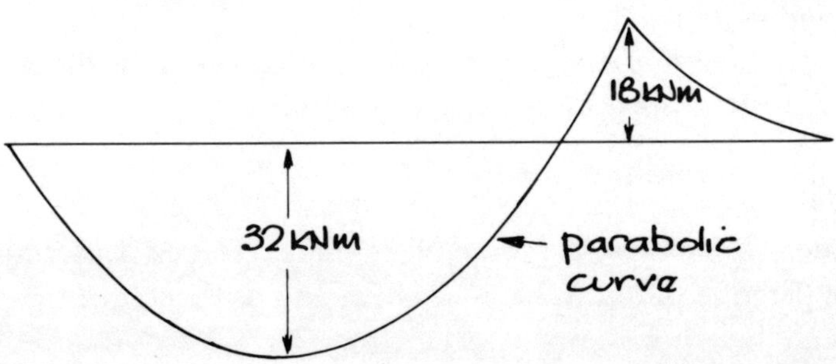

Bending moment diagram

Fig 5.25 Beam with uniform distributed load

have been produced on one figure (fig. 5.25) and incorporate all the work covered in graphical form.

For our final example we can look back at the beam covered in example 5.9 (fig. 5.15), and work through to the B.M. diagram.

Example 5.16

Determine the shearing force and bending moment values for the beam mechanism whose reactions we calculated in example 5.9.

As in example 5.14, the point load forces produce a rectangular form shearing force diagram, starting left to right, i.e. down 40 kN, along 2.5 m, up 168.75 kN, along 1.5 m; down 110 kN, along 4 m, down 110 kN, along 2.5 m; up 131.25 kN, along 1.5 m, down 40 kN (see fig. 5.26). We cross the baseline three times so there are three maximum bending values.

Summating the shearing force diagram from the right:

B.M. at $R_2 = 40 \times 1.5 = 60$ kNm (above baseline)
B.M. at 110 kN load 2.5 m from $R_2 = (91.25 \times 2.5) - 60 = 228.125 - 60 = 168.125$ kNm (below baseline)
B.M. at next 110 kN load $= 168.125 - (18.75 \times 4) = 168.125 - 75 = 93.125$ kNm (below baseline)
B.M. at $R_1 = (128.75 \times 1.5) - 93.125 = 193.125 - 93.125 = 100$ kNm (above baseline)
B.M. at end $= (40 \times 2.5) - 100 = 100 - 100 = 0$

The bending moment diagram follows a triangulated form as for all point-loaded systems.

This completes the study of structural mechanics at this stage but the subject is developed more fully in Volume 2.

Exercises

5.1 Figure 5.27 shows a group of co-planar forces. Determine the resistance force required to balance them and its direction.

5.2 Test a material sample in a laboratory (under supervision) to establish its force extension curve. From this assess its modulus of elasticity.

5.3 Explain the difference between: a force triangle, a force parallelogram, and a force polygon, giving examples of each and their application.

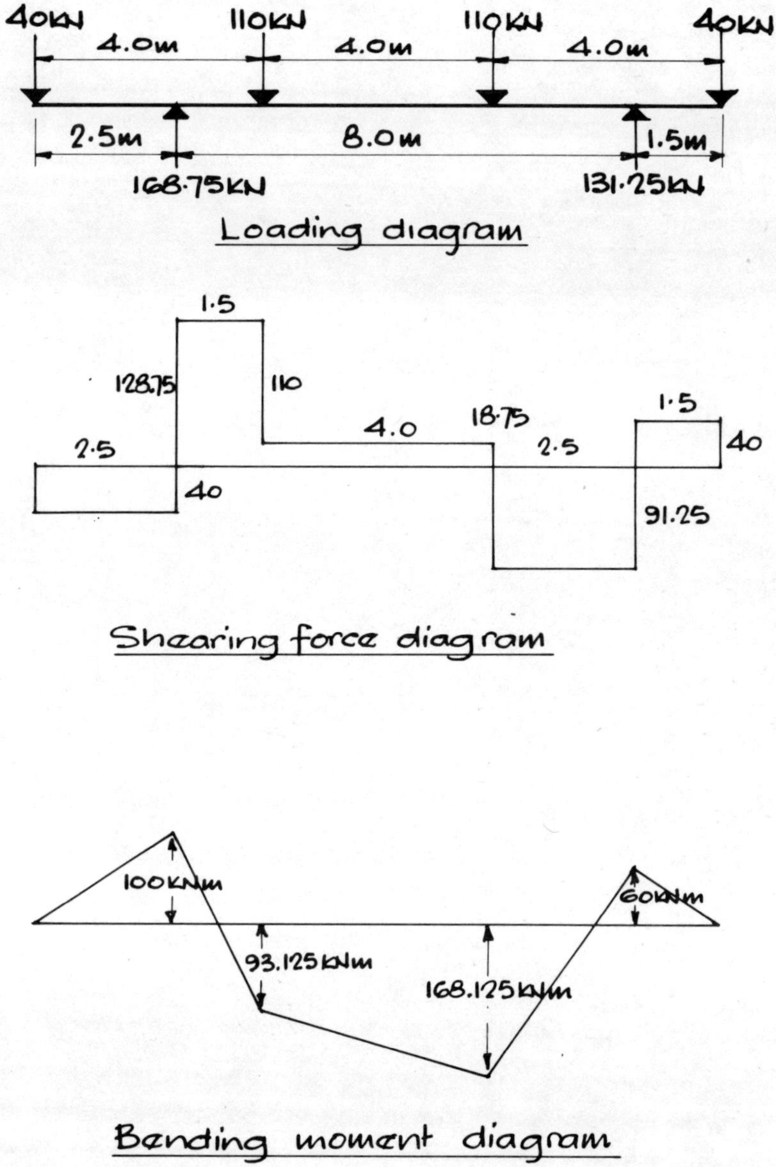

Fig 5.26 Beam with overhanging ends

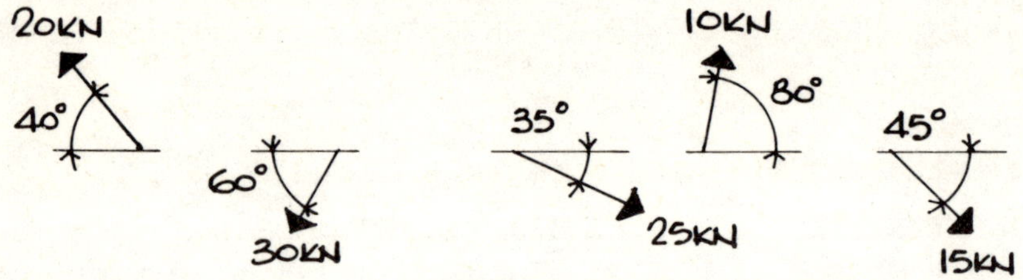

Fig 5.27

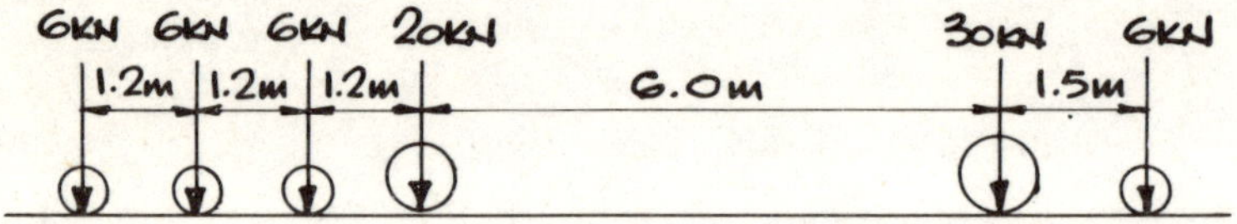

Fig 5.28

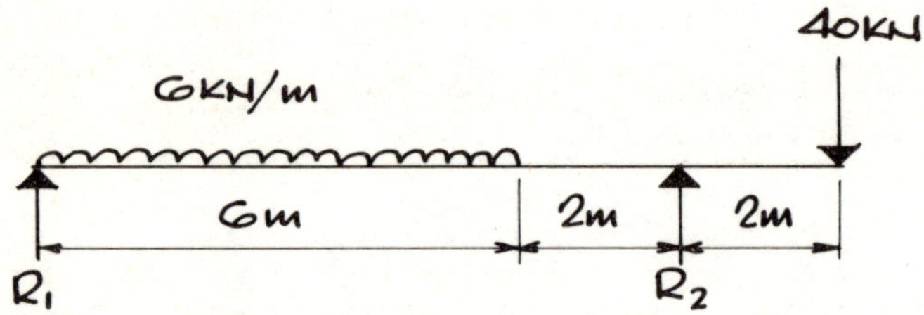

Fig 5.29

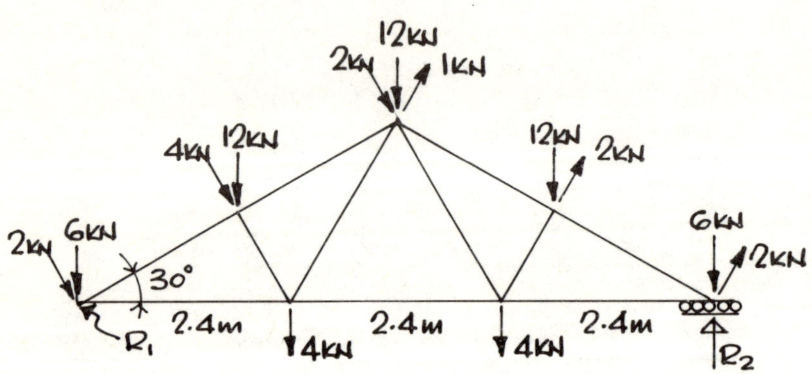

Fig 5.30

5.4 Determine the centre of gravity for the wheel load system whose forces are shown in fig. 5.28.

5.5 Determine the reactions, shearing force and bending moment values for the beam mechanism shown in fig. 5.29.

Assignment

Figure 5.30 shows the loading diagram, including the effect of wind forces, for a timber roof truss. The trusses are at 2.4 m centres and carry purlins and cross-beams to support common rafters and ceiling joists.

(a) Calculate the reaction forces by taking moments about R_1 to find R_2 (assumed vertical force by provision of roller bearing).
(b) Resolve the inclined forces into their vertical and horizontal components to find R_1 and its direction.
(c) Find the resultant forces at the node points to produce the force line.
(d) Construct the force diagram and determine the magnitude and nature of the forces in the members.
(e) Given a rafter section of 125 x 63 mm and a bottom joist section of 100 x 63 mm, determine the changes in length caused by the compression or tension forces, $E = 8\ \text{kN/mm}^2$.

III Environment

6 Thermal Studies

Despite the resources of coal, gas and oil that the U.K. enjoys, attention has been focussed increasingly on the need to conserve energy. The momentum for this was provided chiefly by the oil crises of the early 1970s and the problem shows no signs of abating. Some people would say that this 'energy consciousness' is long overdue.

There are many good reasons for the efficient use of fuels, but as students of construction you should be able to appreciate the benefits to the householder, in terms of smaller heating bills and increased standards of comfort, and to the builder, who can reduce radiator and boiler sizes by providing improved thermal insulation.

HEAT LOSSES FROM BUILDINGS

A large proportion of domestic, industrial and commercial energy consumption is used in space heating (about two-thirds in a typical dwelling). A significant proportion of the remainder goes towards hot water requirements but, because these vary so much according to building type and occupation, we shall not concern ourselves too much with energy consumed in this way. The prime aim of this chapter is to investigate the basic principles governing heat losses from buildings. A closely related subject is the behaviour of water vapour and we shall develop the concepts of condensation from the introduction given in chapter 1, looking at how and why it occurs, its effects on the thermal properties of the fabric, and what provisions need to be made to minimise its effects.

Let us remind ourselves of the ways by which heat energy is transferred. These were mentioned in chapter 4 and are illustrated in fig. 6.1. They are:

(a) *Conduction:* This is the transfer of energy from one molecule or atom to its neighbours. Because this depends on the proximity of the particles, dense materials are usually good conductors.

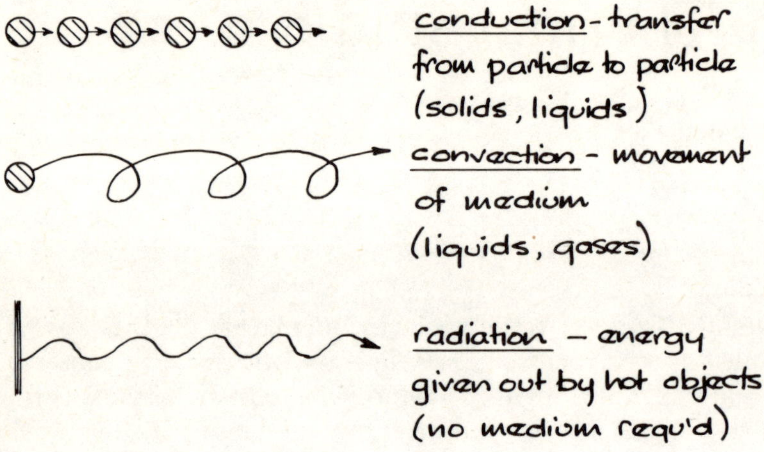

Fig 6.1 Forms of heat energy transfer

Metals, in particular, transmit heat effectively because of the role that 'free' electrons play in energy transfer.

(b) *Convection:* Because fluids become less dense when heated, they tend to rise and are replaced by cooler material. This is also true of air, and convection currents can be seen using coloured dust particles. This principle is responsible for the circulation of air in rooms heated by open or gas fires, by ducted warm air heating, and also by skirting radiators (using finned pipes) since these work mostly by convection. Early forms of solid fuel fired gravity central heating systems also used this principle in the circulation of the hot water and, although modern radiator systems use pumped circulation, radiators on the upper floor will still get warm with the pump switched off. Therefore heat energy is transferred by movement of the medium.

(c) *Radiation:* All surfaces give out electromagnetic radiation. The amount of energy given out, and also its wavelength, depend on the temperature of the surface and its nature. For example, a matt, dark surface (such as a deep-brown concrete tile) will radiate more energy for any given temperature than a light-coloured shiny surface (such as a glazed white kitchen tile). The radiation given off by these warm surfaces is absorbed by other surfaces and it is this effect that makes you aware of heat from, say, a panel radiator of a heated ceiling. Temperatures of wall surfaces in rooms are also important in the transfer of radiant energy from and to occupants of those rooms.

All these forms of energy transfer are involved in the heat loss through the fabric of a building. This is illustrated in fig. 6.2 using a conventional cavity wall as an example.

In order to determine the extent of these losses, for any form of construction, we need to define a property of that constructional element that incorporates all the methods of transfer involved. In reducing the losses, the ability of the solid materials to conduct heat is obviously of importance but we must not neglect the role played by air spaces such as those in cavity walling and double glazing systems. Additionally, at any surface where there is a change in the way that heat is transferred (e.g. from convective to conductive transfer), resistance to overall heat flow is increased and this 'surface resistance' contributes substantially to the overall resistance to heat flow. A good example of this is in a single-glazed window, where the glass itself is a relatively good conductor of heat. Figure 6.3 shows a slab of a building element of area 1 m^2. Let us imagine that the

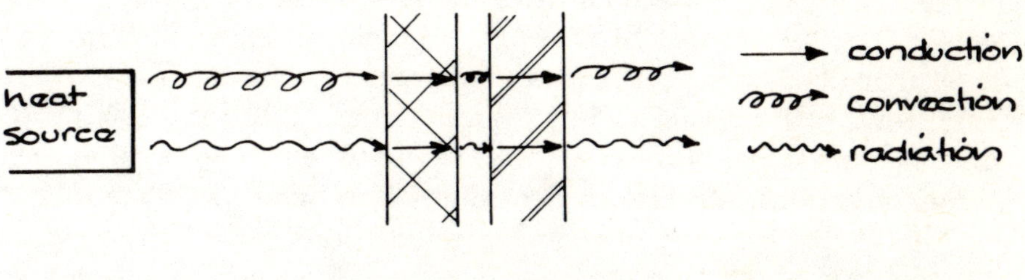

Fig 6.2 Heat transfer through a cavity wall

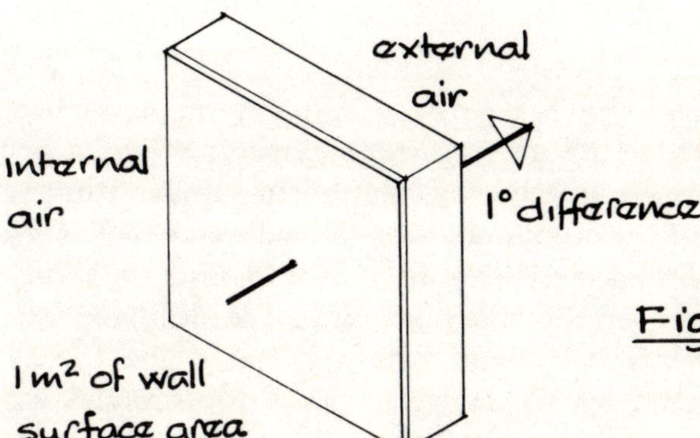

Fig 6.3 Thermal transmittance

temperature difference, measured air to air, is 1°C. The thermal transmittance of the element is defined as the rate of transfer of heat (watts) through the unit area (1 m^2) and for unit temperature difference (1°C) across it. This is known as the U value of the element, which is measured in w/m^2°C. Its size is thus a measure of the effectiveness of the element as a thermal insulator: the lower the U value, the more effective the insulation.

The section of the Building Regulations concerned with thermal insulation is continually being updated, with the effect of reducing statutory U values in domestic construction, and, as techniques and materials are refined, it is likely that the values will be reduced still more.

We can see, then, that the overall heat loss across a wall, roof or floor will depend on its U value, the area through which the heat is flowing and the temperature difference between the internal and external environment. For example, a mid-terraced house will lose less heat than a detached house of the same size because it has only two external walls. Similarly, a bungalow will lose more heat than a house of the same overall floor area because its ground floor and roof areas are greater. In general, the rate of heat flow will thus be proportional to these three quantities and is expressed as $Q = \mathrm{U} \times A \times T$, where Q = rate of heat flow (W), U = U value of element (W/m^2°C), A = area of element (m^2), T = temperature difference (°C). Therefore, for a wall measuring 5 m x 4 m, with a U value of 0.6 W/m^2°C, and a temperature difference of 20°C between its internal and external elevations, the rate of flow Q = 0.6 W/m^2°C x 20 m^2 x 20°C = 240 W. By carrying out similar calculations for all the external elements of a building (or ‘party’ elements where applicable), an estimate can be made of overall ‘fabric’ losses.

Example 6.1

Let us consider the detached house, illustrated in fig. 1.12 of *Building and Civil Engineering Construction* Volume 1, to determine the heat losses through the fabric in 1 h for a consistent temperature difference of 16°C. It is useful to set out the calculation in tabular form to avoid missing out any of the components or elements (see table 6.1). From this it will be seen that the total rate of heat loss is 3.68 kWh.

You will notice that, for this example, since the heat loss was consistent for all elements, we could have summated the $U \times A$ values and *then* multiplied by 16. However, since these values are not always consistent, e.g. the temperature difference of party walls

Table 6.1 Values used in example 6.1

Element	U value (W/m^2°C)	Area (m^2)	Temperature difference (°C)	Loss (Q) (W)
Walls	0.60	133	16	1277
Roof	0.35	64	16	358
Floor	0.55	64	16	563
Doors and windows	2.80	33	16	1478
Total				3676

would probably not exceed 4°C, and ceilings of flats might also be about this temperature difference, it is always safer to include the T value in the tabulation. Since we are investigating a 'new' property, it must conform to current Building Regulation requirements, the appropriate part of which may be summarised as follows:

Element of building	*Maximum permitted U value*
External wall	0.60 W/m^2°C
Floor	0.60 W/m^2°C
Roof	0.35 W/m^2°C

Also, the area of double-glazed windows should not exceed 24% of the external wall area (12% for single-glazed windows). The restriction on window area is very important and this can be seen in the contribution to the total heat loss for the example shown, the doors and windows contributing 1.16 times as much as the walls even though they form only 0.2 of the overall wall and window area.

[*Note:* At present the regulations allow an increase in window area if the wall has a U value better than 0.6 W/m^2°C, e.g. if the U value of the wall were 0.4 W/m^2°C (which is easily possible in timber-framed construction), the permissible area of double-glazed windows would rise to 30% of the total wall/window area. This would seem illogical from a 'design' consideration, since part of the reason for improving the wall value is to improve the performance over all the building, not to allow the extra heat loss to be produced elsewhere! In practice the window area contribution is often reduced *below* the permitted percentage (as in this example, 20% instead of 24%) to improve performance.]

Approximate contributions to overall heat loss from a typical dwelling are illustrated in fig. 6.4. The values are for current *U* value requirements. The contributions for an older property (say 1920) with no insulation are shown in brackets. The overall heat loss for the latter is, however, some 2.5 times that of the modern well-insulated house. It does show that the roof space is the area that is most easily improved and in part the Building Regulations requirements reflect this as being achieved both easily and cheaply. The figures shown are only approximate and refer to a detached property. The type of property, its location and the level of insulation will produce variations. A terraced house with similar insulation would, for example, lose a greater proportion through the roof because the wall/window area is considerably reduced. Its overall heat loss figure would, however, be much smaller.

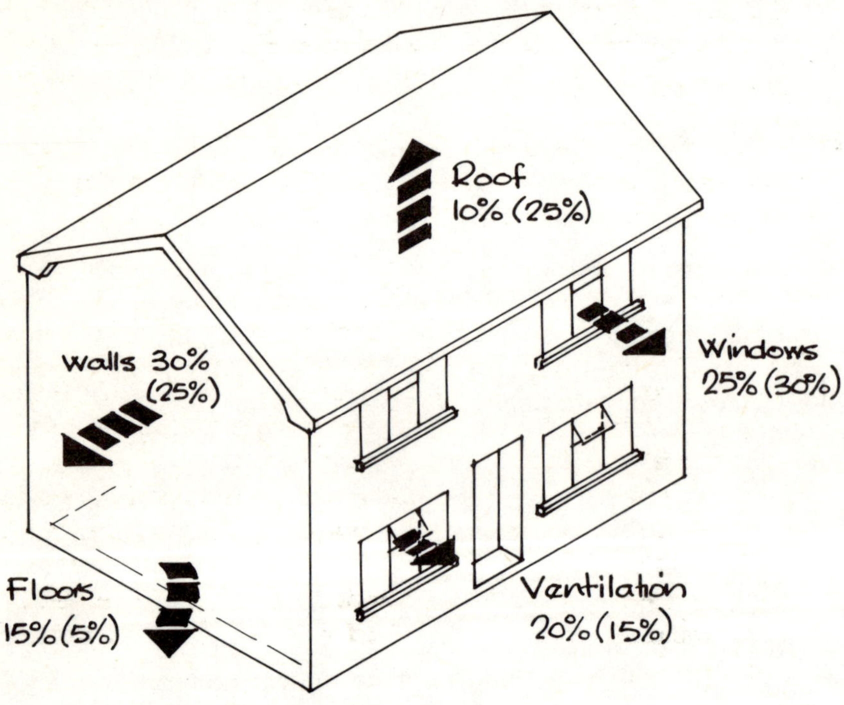

Fig 6.4 Heat losses, in proportion, from a typical detached house

From fig. 6.4 we can see that ventilation losses account for approximately one-fifth of the space heating costs. Ventilation of buildings is necessary for a number of reasons, as follows:

(a) provision of oxygen for respiration of occupants and combustion of gas, coal or oil;
(b) removal of waste gases, contaminants and odours;
(c) reduction of condensation;
(d) prevention of fungal and mould growth.

Because we are concerned with air temperatures, for comfort conditions, the incoming air must be heated to the desired internal temperature of the building. The rate at which energy is expended on this depends on three factors:

(a) the temperature interval through which the air is heated.
(b) the rate of ventilation, expressed in air changes per hour, which includes calculation of the volume of air required to be changed. This will depend on the use of the building; indeed it will vary for different zones of the building. For detailed requirements you should refer to the Chartered Institute of Building Services (C.I.B.S.) Guide.
(c) the specific thermal capacity of air. This is the quantity of heat required to raise the temperature of 1 m^3 of air by 1°C, and is normally taken to be 1.2 kJ.

Example 6.2
Let us consider the house used in example 6.1 to determine the loss from ventilation. The volume is approximately 320 m^3 and an *average* ventilation rate of 0.8 air changes per hour is assumed. This takes account of kitchen and bathroom requirements which, although high, are occasional. The total volume of air that requires heating for 1 h therefore equals $320 \times 0.8 = 256\ m^3$. In 1 h, therefore, the amount of heat needed (for a temperature change of 16°C) is given by $256 \times 1.2 \times 16 = 4915$ kJ.

Since we need an answer in kilowatt hours, we can refer to chapter 4 where we stated that 3.6×10^3 kJ = 1 kWh. The rate of heat loss will be:

$$\frac{256\ m^3 \times 1.2\ kJ/m^3\,°C \times 16}{3.6 \times 10^3} = 1.37\ kWh$$

This represents (air volume × specific thermal capacity × temperature rise)/3600. If we now summate the fabric losses, and those caused by air changes, the total heat loss equals 3.68 + 1.37 = 5.05 kWh.

We are now part way towards finding out how much it will cost to heat the house but, before we can be definite, it must be stressed that our calculations are heavily dependent on the temperature difference we use. There may not be a constant temperature requirement throughout the house. During the day, bedroom temperatures may be lower than those of living rooms and we may make the assumption that, if bedrooms are at 16°C and living areas are at 20°C, the average internal temperature will be 18°C.

External temperatures will obviously vary throughout the year but we shall concern ourselves with peak loads that are likely to occur in the winter quarter taking in the months of December, January and February. Detailed information on U.K. outside temperatures is readily available from the C.I.B.S. Guide and the calculation uses an average external temperature of 2°C for the period concerned, this being typical of the Midlands. If we compare this with an external temperature of −10°C for the same period in Ottawa, Canada (giving a temperature difference of 28°C rather than 16°C), it is easy to see why standards of thermal insulation in such countries have been higher than those in the U.K. Additionally, the *U* values we have adopted are based on *normal* conditions of exposure. Since external surface resistances vary considerably according to the type of location, *U* values may need to be modified for sheltered conditions or those of severe exposure, such as are found on coastal sites. In particular, *U* values for single-glazing are very variable, ranging from 5.0 $W/m^2 °C$ for sheltered exposure to 6.7 $W/m^2 °C$ for severe exposure, giving an increase of 34% on heat losses. The difference becomes smaller for more effectively insulated structures, the corresponding increase for double-glazed equivalents being little more than 10%.

It may be useful at this stage to compare our new house, which meets current Building Regulations, with one of similar dimensions which was built in the 1950s with no statutory requirements for insulation. Such a house would have two skins of brickwork with a ventilated cavity, single-glazed windows, a solid ground floor finished with thermoplastic tiles and a plasterboard ceiling (but no insulation) to the loft space. The fabric losses, still using a 16°C temperature difference, would be as indicated in table 6.2, with the earlier figures shown in brackets for comparison.

Table 6.2 Fabric losses from a 1950s house

Element	U value ($W/m^2\,°C$)	Area (m^2)	Temperature difference (°C)	Loss (Q) (W)
Walls	1.48 (0.60)	133	16	3149 (1277)
Roof	2.60 (0.35)	64	16	2662 (358)
Floor	0.75 (0.55)	64	16	768 (533)
Doors and windows	5.70 (2.80)	33	16	3010 (1478)
Total				9589 (3676)

It is also likely that the lesser regard to draughtproofing of doors and windows in the older house would produce significantly greater ventilation losses. If we make a conservative estimate and say that the ventilation rate increases from 0.8 to 1.25 air changes per hour, the ventilation losses would increase from 1.37 kW to 2.14 kW. Total losses will thus be 5.05 kWh for the new house and 11.73 kWh for the old house.

Example 6.3

Before looking at how the losses have been reduced, we should calculate the heating costs for the two houses for a typical winter quarter. As explained in chapter 4, we use kilowatt hours rather than joules since the values produced are more manageable numerically. Also, since the heat losses have been calculated for a *1 h* period with the value in *kilowatts*, it is logical to use kilowatt hours. Thus in order to cost the heating we multiply 1 h operation by the number of hours that the heating is being used. Very few households operate the central heating system 'around the clock' and typical weekday times might be 06.00 to 10.00 and 16.30 to 22.30, giving 10 h per day. If we allow an extra 4 h per day at weekends, this produces a weekly figure of 78 h. The 13 week quarter means 78 x 13 = 1014 h, which for calculation purposes we will approximate to 1000 h.

The energy used for the new house is therefore 5.05 x 1000 = 5050 kWh, which compares with 11730 kWh for the old house. If our energy is costed at 4p per kilowatt hour (an arbitrary value), the cost would be 5050 x £0.04 = £202 for the new house as opposed to about £470 for the poorly insulated older house. This represents a saving of £268 for *one* winter quarter, and the saving could probably exceed £500 for a full year. It should be borne in mind, lest you start comparing heating costs in your own homes, that we have not included domestic hot water, cooking costs or the running costs of appliances such as washing machines and dishwashers. However, none of these is affected by the insulation of the building fabric although it would be sensible to insulate the hot water storage cylinder to conserve energy, as explained in chapter 4.

Having shown that a saving in cost can be effected by good thermal insulation, let us consider some of the ways in which the older house might be upgraded, or the new house construction might differ. For a new house it is likely that some less traditional materials might be used in its external structure whereas with the older house the insulation will need to be applied, preferably with a minimum of disturbance to the building fabric. The Building Regulations allow plenty of scope with regard to structural form and also to the selection and thicknesses of insulating materials which are 'deemed to satisfy' the standards. The *U* values of many constructional forms and insulation systems can also be found in the C.I.B.S. Guide.

The calculation of *U* values from first principles requires a knowledge of the thermal conductivities of the materials used. We have discussed how air spaces and surface resistances are also important in establishing overall *U* values and a more detailed consideration will be found in Volume 2. It is sufficient at this stage to use 'given' *U* values in your exercises and assignments and we shall restrict our discussions to the general properties of insulators and insulating systems and how they can be used in buildings.

Provision to meet requirements is obviously much easier to achieve when building 'from scratch' and typical forms of domestic and commercial construction are shown in *Building and Civil Engineering Construction* Volume 1. The upgrading of existing elements of construction is more complex, however, and we shall concern ourselves with this. Let us consider the more common methods of reducing fabric losses in these existing buildings and their constructional elements.

WALLS

(a) Dry lining of internal walls by plasterboard, either on battens or plaster dabs, the former allowing the use of polystyrene-backed boarding. This would produce a reduction from 1.48 to 1.1

$W/m^2\,°C$ (plasterboard) or 0.57 $W/m^2\,°C$ (with expanded polystyrene backing).

(b) Cladding or rendering the outside walls using expanded lightweight materials. This could effect similar savings in heat loss but might pose problems on impact resistance allied to weathering ability.

(c) Filling the wall cavity with urea–formaldehyde foam, expanded polystyrene beads or blown rock fibre. For any form of building other than detached it will be necessary to convince occupants of the attached premises that they should also invest in 'cavity fill' since if one part alone of the building were insulated, the cavities to the adjacent walling would be 'bridged' and create dampness in the walls! The success of the filling is very much dependent on the quality and supervision of the workmanship. The filling must be *complete* to avoid tracking of damp across the cavity and, in the case of the foam, the work must be allowed to dry fully. It does, however, have one distinct advantage in that the inner leaf of the wall is warm and unlikely to be subject to interstitial condensation. For the wall with two half-brick skins, the U value is reduced from 1.48 to 0.75 $W/m^2\,°C$.

[*Note:* In method (a) there would need to be some modification to window surrounds and window boards, and internal fixings would be more difficult to achieve; whereas in method (b) the window cills would probably need replacing in order to retain effective weather grooves.]

ROOF ZONES

(a) Loose-fill insulation between the ceiling joist sections of the roof structure. This is easy to apply, producing its own width between timbers, and thickness depends on the volume of material used. The insulation values are much the same as for method (b) which follows, but if the roof space is well ventilated the vermiculite granules can be blown away from the eaves, producing a 'cold bridge'.

(b) Glass fibre quilting between the ceiling joist sections of the roof structure. This is fine for spacings of 400 or 600 mm since rolls of these widths are available. For other spacings, however, the roll will either overlap the space (no real problem) or underfill the space (causing strips of cold bridging). Particular attention needs to be paid to the eaves detail to avoid either cold bridging or lack of ventilation to the roof space, producing good conditions for the promotion of dry rot!

The reduction in *U* values is dependent on the thickness of the insulation but this does not produce a linear reduction, inasmuch as a 100 mm thickness is not four times as good as a 25 mm thickness. Values above 150 mm have not, therefore, been included in table 6.3. These figures compare with the original of 2.60 W/m^2°C, and even a 25 mm thickness shows a considerable saving and 150 mm shows a reduction in heat loss of 91%. Since this is the cheapest form of thermal insulation, it will pay for itself very quickly.

Table 6.3 Thicknesses of insulation to reduce *U* values in pitched roofs

Thickness of insulation (mm)	*U* value (W/m^2°C)
25	0.99
50	0.61
75	0.44
100	0.35
150	0.24

WINDOWS

(a) Replacement of the glass by 'sealed unit' panes with either a vacuum 2–3 mm space or a 6 mm hermetically sealed gas-filled space. Not available on all pane sizes.
(b) Replacement double glazing using sealed units of either 12 mm or 20 mm gap. This means removing existing windows and sealing the perimeters of the frames replaced. Units are often 'made to measure'.
(c) Secondary double glazing providing a gap of between 20 and 60 mm. The units may not always blend easily with the existing construction but are available in hinged or sliding form (the latter being cheaper).

The *U* values vary according to the amount of window area taken up by the frame, the material of the frame (and possibly its insulation), the type of glass (a reflective glass has been developed for use in double-glazed sealed units) and to some extent the space between panes. Table 6.4 shows an approximate set of *U* values for average conditions. As can be seen, the effect of double glazing will reduce heat loss through the windows. It is, however, very costly if considering replacement windows of good quality design and construction. Also, it should be remembered that windows are a useful means of escape in the event of fire. Some cheap fixed-pane windows with very small top opening fanlights remove this means of escape. It is far better if buying or specifying replacement windows to ensure that their style is in harmony with the general design of the building and that they do allow exit in case of fire. In the case of secondary double glazing, 20 mm is probably the optimum space between glass panes since convection could occur at spaces slightly greater than this. All systems should reduce draughts and noise.

Table 6.4 Effect of double glazing in reducing *U* values from 5.7 W/m^2°C

Type of window or gap between panes	*U* value for average location and exposure (W/m^2°C)
2–3 mm vacuum gap and 6 mm sealed unit	3.6 – 3.3
12 mm sealed unit	3.3 – 3.0
20 mm sealed unit	3.0 – 2.8
20 mm sealed unit with treated glass	2.5 – 2.3
20 mm and above secondary glazing	2.8

FLOOR SLAB

The provision of insulation to an existing solid concrete ground floor slab is not normally a practical proposition since it affects doors, fitted units and thresholds at entrances to the house. Some insulation can, however, be provided by fitting mosaic wood block flooring panels. However, the rate of heat loss through the ground slab is not as high as in the remainder of the external fabric since the ground itself does not lose heat at the same rate as the external air. The losses at the perimeter of the slab are more significant since the heat passage is assisted by conduction through the soil to the atmosphere. This is one of the cold bridge zones discussed in chapter 1 and the best solution to the problem in an existing property is to install skirting radiators. For new work the insulation is best sited below the oversite for about 1 m in from the edge of the slab–wall intersection but floating floors are an alternative.

PROPERTIES OF INSULATION MATERIALS

Having considered methods by which buildings may be insulated, let us examine the features of the materials used.

Low thermal conductivity

Figure 6.5 shows a simple method of assessing thermal conductivity for various insulating materials. The tube is filled with hot water and a time–temperature graph is produced to record its rate of cooling. By comparing the cooling curves for each material, a qualitative measure of their conductivities can be made. A more precise method, using Lee's disc, is described in Volume 2.

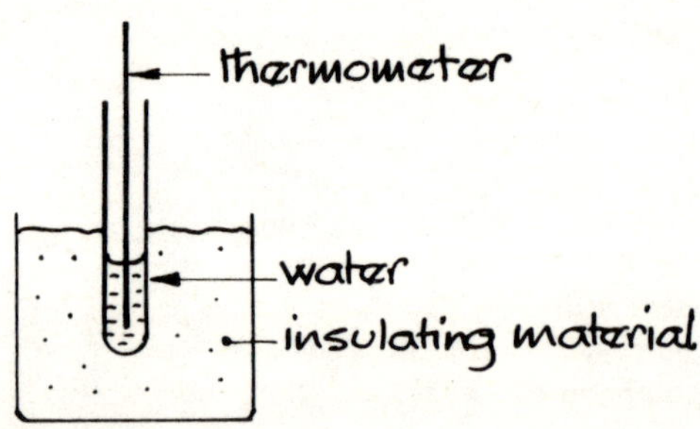

Fig 6.5 Effectiveness of insulators

Still air

It has already been seen that the trapped air in expanded polystyrene, glass-fibre quilting and lightweight concrete blocks is a poor conductor of heat. In the same way, the air gap in double glazing provides much more resistance to heat flow than the glass. The spacing of the panes is, however, important, as is seen in table 6.4.

Reflection of heat

Bright metals such as aluminium foil reflect heat and can be used to effect as a backing to plasterboard on dry-lined external walls, or as a backing to sarking felt under roof tiles. They may also be used behind radiators but in this context are likely to lose their efficiency by collecting dust and grime from the atmosphere through a mixture of convection and water vapour in the air.

It should be appreciated that not all forms of additional insulation are equally effective and consideration should be given to a comparison of annual savings balanced against installation costs (including interest on capital borrowed). This will be seen in developing the assignment at the end of this chapter.

We talked in chapter 1 about 'cold bridges', and mentioned them again in this chapter, so what exactly is a 'cold bridge'? When the *U* value of the external fabric changes abruptly, by a change of material, by a loss of insulation through bad practice or poor detail, or because of failure to upgrade the insulation to compensate for the material change, the internal surface locally will be much colder than that adjacent to it. As an example, if the two lintels shown in fig. 3.5 of *Building and Civil Engineering Construction* Volume 1 are compared, the pressed steel lintel with added thermal insulation has a *U* value close to that of the wall and in fact provides a similar internal surface by inserting a lightweight block in the channel whereas the prestressed concrete lintel is a solid dense material which continues the full thickness of the external wall. This results in *U* values of 0.6 $W/m^2 {}^\circ C$ for the pressed steel installation and 2.7 $W/m^2 {}^\circ C$ for the concrete lintel. For an internal temperature of 18°C, the surface temperatures would be 17.8°C and 12.8°C, respectively, showing an acceleration of local heat loss and a risk of condensation.

We have already talked about condensation and ventilation but in considering 'controlled' ventilation let us examine the measures that can be adopted to reduce indiscriminate air changes. In addition to insulating the extruded frames, manufacturers of aluminium window frames incorporate either a nylon pile sealing strip (for sliding windows) or a Neoprene weather strip (for side- or top-opening windows) to exclude draughts. These have a secondary effect in that they separate the frames and produce a *thermal break*, which further reduces the risk of cold bridging. Some external doors incorporate either impact draught seals or magnetic seals, which are located on the frame and the door. A third important area is in the provision of lobbies between the entrance doors and the inner halls. This can limit the amount of warm air lost directly to the external atmosphere when opening the door, by comparison with a traditional design where warm air from both the hall and the stairwell is lost, this being a much greater volume of air which then has to be reheated.

In discussing heat losses we should consider an alternative approach to the economic use of energy, which is control of the heat output. We touched on this in example 6.3 when talking about operational time. Most modern central heating systems have some forms of control to reduce running costs and the more efficient of these include:

(a) *Programmer:* this can vary from a simple clock timer, which will switch the boiler on and off twice in a 24 h cycle to a full set of programmes that select central heating and hot water in various combinations of operation.
(b) *Boiler thermostat:* this will regulate the operating temperature of the water being pushed round the system, including the heat exchange to produce domestic hot water.
(c) *Hot water thermostat:* which will operate a cut-out on the heat exchange.
(d) *Room thermostat:* which can set the temperature to an acceptable level. By careful location this can vary the output for bedrooms and living area and in split systems can be provided for each zone. By dropping the required temperature from 20°C to 19°C, a large saving in costs (6% on an outside temperature of 2°C) can be made.
(e) *Radiator thermostats:* these act as heat-sensitive valves which shut flow through the radiators when heat levels are sufficient. Fitting these to each radiator on the system can probably reduce costs of heating by up to 15%. Manual control is, of course, possible, even without such thermostatic valves, since normal valves can be either switched off or regulated to reduce flow.

MOISTURE IN BUILDINGS

In chapter 1 we looked at causes of water in vapour form and to appreciate the problems we need to understand how evaporation of liquid water occurs. In any liquid the molecules move with random velocities, as shown in fig. 6.6, and at greater temperatures the the average velocity will increase. The surface effect of the liquid is such that a molecule can only pass through this barrier, into the vapour state, when it has sufficient energy. Since it is only the higher energy molecules that evaporate, the average velocity of those re-

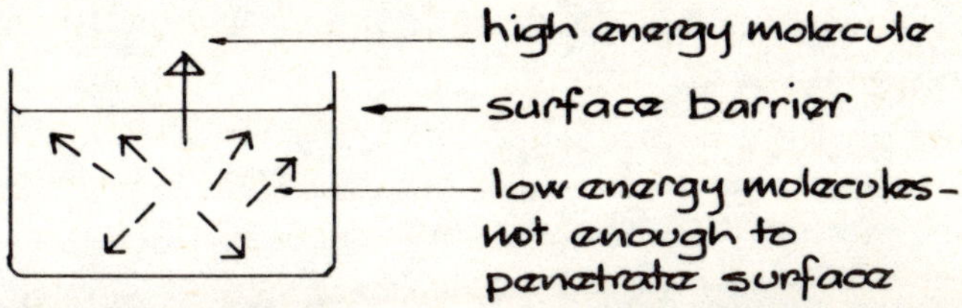

Fig 6.6 Evaporation of water

maining is reduced, which results in a cooling effect. Once in the vapour state, these molecules exert a vapour pressure by collision with the walls of the vessel shown. At any given temperature, for a closed (unventilated) system, the space may become saturated and unable to accept more vapour. The pressure then exerted is the saturated vapour pressure (s.v.p.), as shown in fig. 6.7. Because more molecules have sufficient energy at higher temperature to enter the vapour state, saturated vapour pressure increases with temperature. In simple terms, this means, as stated in chapter 1, that warm air can hold more water vapour than cold air. When warm, moist air is cooled, it follows that a temperature may be reached when the air cannot hold the water in vapour form, at which point condensation occurs. This temperature is called the *dew point*, which does not have a fixed value but is dependent on the air temperature and the amount of water vapour present. We may describe the amount of vapour present in terms of relative humidity (R.H.), i.e.

$$\text{R.H.\%} = \frac{\text{Mass of water vapour in air}}{\text{Mass it could hold if saturated, at that temperature}} \times 100$$

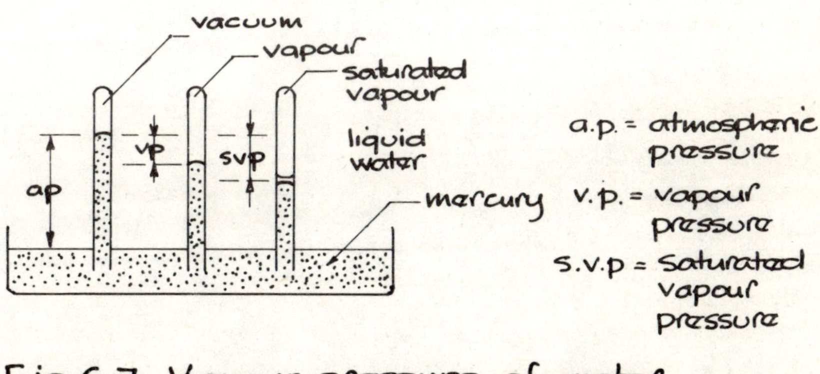

Fig 6.7 Vapour pressure of water

If we then remember that the mass of vapour is directly proportional to the pressure it exerts, then

$$\text{R.H.\%} = \frac{\text{Vapour pressure}}{\text{Saturated vapour pressure}} \times 100$$

The relative humidity, and hence the dew point, may be found by using either a paper hygrometer or a *sling hygrometer*, as illustrated in fig. 6.8. The sling hygrometer works by comparing the cooling effect of the atmosphere, on the wet bulb, with the air temperature. Ease of evaporation, and hence the difference in temperatures between the wet and dry bulbs, are dependent on a high air temperature and low moisture content of the atmosphere. Relative humidity is obtained either by using tables supplied with the hygrometer or from a psychrometric chart whose use is described in Volume 2. The paper hygrometer depends on the expansion of cellulose-based materials caused by increasing moisture content and is not as accurate.

Since the effect of cold surfaces has already been mentioned, you may like to find relative humidities in a varicty of rooms. For example, a college canteen or a full classroom gives high humidity and it is worth noting both the air and surface temperatures to assess the risk of condensation. If surfaces are below the dew point, then condensation will occur as the warm air is cooled on contact. The dew point may be determined empirically using Regnault's hygrometer (see fig. 6.9), which shows the formation of condensation rather well.

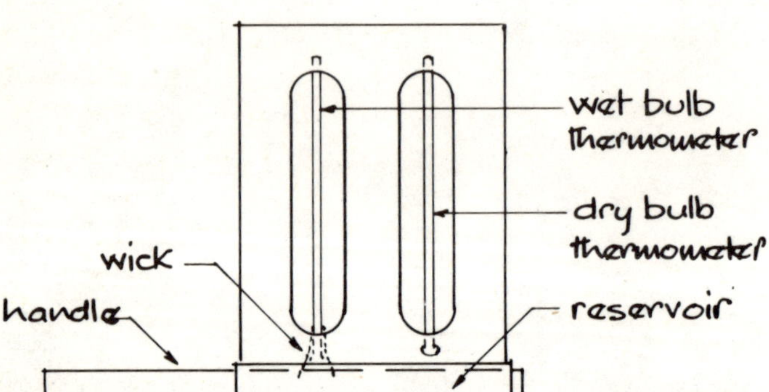

Fig 6.8 The "sling" hygrometer

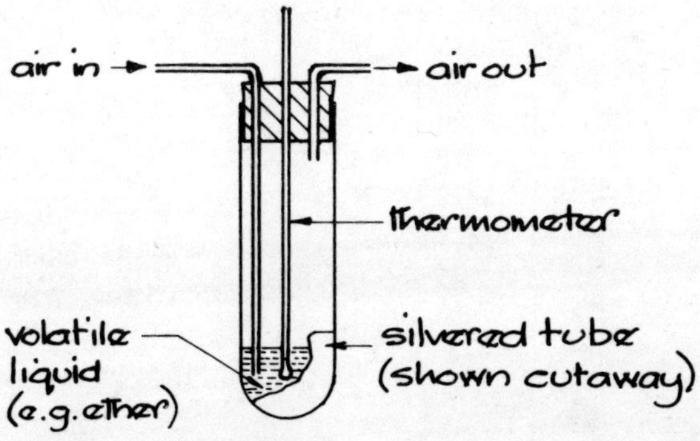

Fig 6.9 Regnault's hygrometer

We have discussed previously the damaging effects that such condensation can cause, and we can now look at steps that may be taken to minimise the problem. These include:

(a) Raising surface temperatures by *increasing general heating levels:* you will probably have noticed that condensation is usually more apparent in colder rooms.
(b) *Providing additional insulation:* if this is applied internally, surface temperatures will rise but fabric temperatures may fall, leading to other problems, discussed later.
(c) *Providing double glazing:* this will only work if there is an effective seal preventing vapour from coming into contact with the outer pane.
(d) *Reducing the quantity of vapour present in the air:* this might be achieved by the installation of mechanical extractor fans. Typical locations are kitchen and laundry rooms. It is not usual to install them in bathrooms (except when they have no windows) since they are not normally used for as great a length of time as kitchens, and sufficient ventilation may be provided by opening windows for a short time.

We have seen that many building materials are porous and, because water vapour pressure inside buildings is greater than that outside, the air containing water vapour will be forced into the fabric of the building, as shown in fig. 6.10. If, at any point in the structure, the temperature falls below the dew point, then condensation will

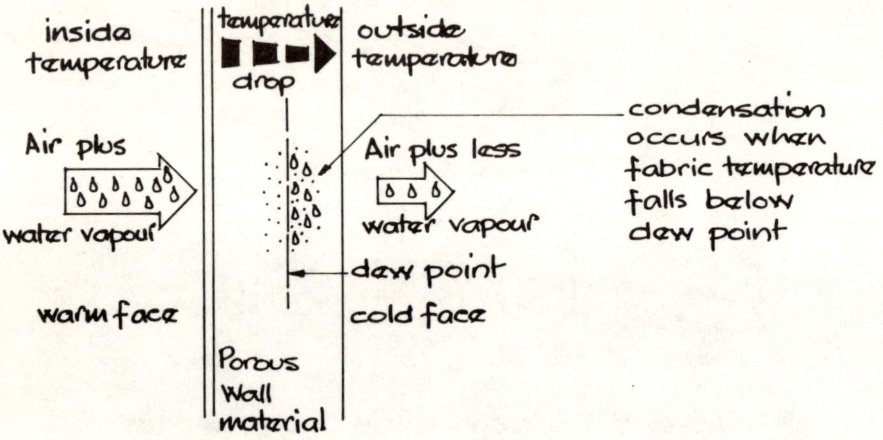

Fig 6.10 Interstitial condensation

occur within the pores of the material. In addition to the problems outlined in chapters 1 and 2, there is a possibility that the insulating materials will become damp and lose their effectiveness, since water is a fairly good conductor of heat.

The control of interstitial condensation involves not only the methods outlined previously but also keeping as much of the structure as possible in a warm state. As mentioned earlier, in the case of upgrading the existing property, it is not realistic to insulate the outside face of a wall even though (from a purely scientific approach) this is the most efficient solution to avoiding interstitial condensation. This is not, however, true of flat roof construction and examples of 'warm roof' construction are shown in *Building and Civil Engineering Construction* Volume 2. These follow the principle of using bitumen-coated lightweight aggregate *above* the felt or asphalt (to a thickness of at least 100 mm) to hold the heat in, retaining heat in the roof deck and thereby minimising the risk of condensation.

In cavity wall construction the insulation is best bonded to the cavity face of the inner leaf, thus retaining a 25 mm cavity as described in chapter 1 but maintaining a 'warm' inner wall.

For timber framing, both in stud walls and timber-joisted flat roofing, the vapour barrier is of vital importance to ensure that dry rot is not allowed to develop. If, in addition to the polythene sheeting, a polystyrene-backed plasterboard is used and sealed on its decorative face prior to further decoration, this will help to improve resistance to the passage of water vapour. The ceiling can also be finished with a textured plastic coating, such as Artex, to form an additional barrier.

Exercises

6.1 Calculate the rate of heat loss through a wall of area 20 m^2 and U value 0.6 W/m^2°C subjected to a temperature difference of 20°C. If the U value is reduced to 0.5 W/m^2°C, by improved insulation, and the internal temperature requirement is reduced by 1°C, determine the revised heat loss.

6.2 A building has a volume of 100 m^3 and experiences 2 air changes per hour. Given that the specific thermal capacity of air is 1.2 kJ/m^3°C, determine the rate of ventilation losses for air heated by 16°C. If the rate of ventilation is controlled to 1.2 air changes per hour and the internal temperature requirement is reduced by 1°C, calculate the revised ventilation losses.

6.3 Using the U values for the component parts of a building as given in example 6.1, and assuming a temperature difference of 15°C with 1.2 air changes per hour, calculate the rate of heat loss for a mid-terraced house of the following critical dimensions:

External walls	42 m^2
Floor	40 m^2
Roof	40 m^2
Doors and windows	12 m^2
Volume	200 m^3

Ignore heat loss through party walls, which are small.

6.4 In the winter quarter the boiler is operational from 06.30 to 08.30 and 17.30 to 22.30 on weekdays and from 08.30 to 12.30 and 15.30 to 22.30 at weekends. Using the figures obtained from exercise 6.3, determine the heating costs (excluding domestic hot water) for the complete quarter at a unit cost of 3.2p per kilowatt hour.

6.5 In the use of the sling hygrometer:

(a) when will the two temperatures be the same?
(b) what does a large difference in temperatures signify?
(c) what would you observe if you used a liquid of lower boiling point in the reservoir?

Assignment

Examine the construction of your own house or flat and assess, as well as you can, the areas of the external elements: walls, roof, floor, windows and doors. From technical data assume approximate *U* values for these elements and determine the fabric heat loss for a temperature difference of 16°C.

Determine the volume of air (to underside of roof space) and, assuming 1.5 air changes per hour and a specific thermal capacity of air as 1.2 $kJ/m^3 °C$, calculate the ventilation losses.

Estimate the number of hours the heating is operational per week, and from this per quarter, and establish the total heat loss in kilowatt hours. Using a current rate of fuel costs, determine the cost for the quarter (exclusive of domestic hot water) and compare your figures with an actual fuel bill. Calculate what the savings would be if:

(a) an additional 75 mm of loft insulation were provided (refer to table 6.1 for approximate figures);

(b) replacement or secondary double glazing were fitted (assume a revised *U* value of 2.8 $W/m^2 °C$).

Estimate the additional costs involved from price lists or, in the case of (b), installers' quotations, and (allowing for interest rates on borrowed capital) calculate the pay-back time needed for each before an *actual* saving is achieved.

7 Electricity

In chapter 4 we talked about heat as a form of energy and illustrated the use of electricity for heating water. Much of the space heating, hot water and (of course) lighting requirements in domestic, commercial and industrial buildings relies on the electricity supply. Also street lighting, traffic signals, generators for pumping plants and a host of mechanical operations depend on electricity. From this we can see that it is a form of energy and that one of its useful properties is that it can be converted easily into other forms.

THE NATURE OF ELECTRICITY

As you will know, atoms consist of positively charged nuclei surrounded by negatively charged electrons. The outer electrons of good *conductors*, such as metals, are not tightly bound to the atoms and, if these electrons can be made to move, an electric current will be produced. In contrast, the electrons of poor conductors, such as plastics, rubber and dry timber, are not so free to move, which is why they make good electrical *insulators*. Halfway between these extremes are materials such as silicon and germanium, which are called *semi-conductors*, and these are used quite extensively in the electronics industry.

Electrical flow is in many ways similar in its behaviour to the flow of water and needs a 'driving force'. This is illustrated graphically in fig. 7.1, which shows a working model of a water circuit. When the tap is closed, no water can flow but the pressure energy caused by the head of water (H) is ready to be converted into energy movement (the potential energy described in chapter 4). In electrical terms this static pressure is called *electromotive force* (e.m.f.), which is measured in volts (V). When the tap is opened, water flows (a current in electrical terms) and there is a pressure drop between the tappings. In electrical terms this pressure drop is called the *potential difference* between the points and is also measured in volts. If we take this comparison

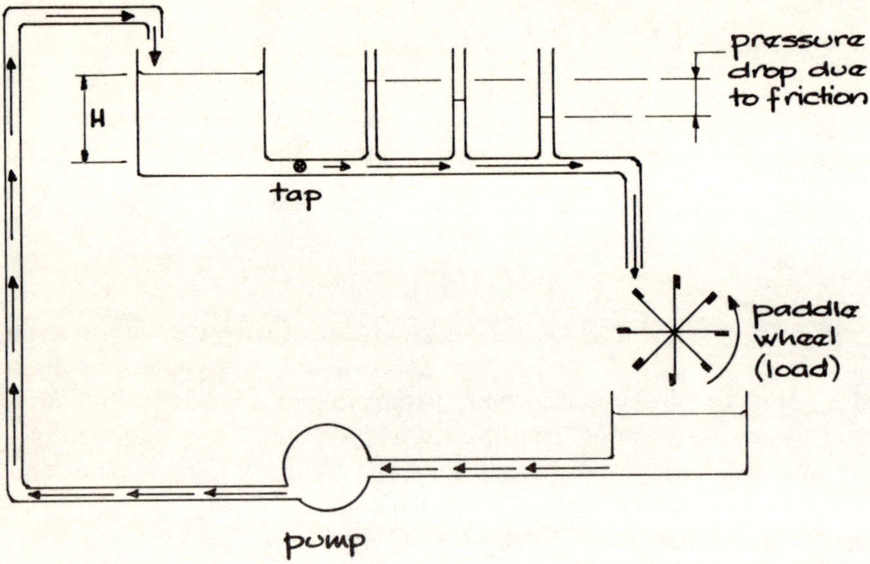

Fig 7.1 Similarity between flow of water and flow of electricity

with water still further, we can see that the volume flow rate between the tappings will depend on both the pressure drop and the resistance created by friction within the system, this being dependent on the length and cross-sectional area of the pipe and the roughness of its surface. In electrical terms the current (I), which is measured in amperes (A) depends on the potential difference (V) and the electrical resistance (R), which is measured in ohms (Ω). This resistance is determined by the length of the conductor, its cross-sectional area and the material from which it is made.

The further apart the tappings are spaced, the greater will be the resistance to flow. The resistance may also be increased by reducing the bore of the pipe. In an electrical circuit, the value of the resistance R of a conductor length l and cross-sectional area a is given by $R = \rho\ (l/a)$ where ρ is the resistivity of the material. Some resistivities are given in table 7.1, which shows that metals, such as copper and brass (which are widely used in circuits), have very low resistivities.

Table 7.1 Material resistivities in ohm metres

Material	Resistivity (ρ)
Copper	1.72×10^{-4}
Brass	8×10^{-4}
Aluminium	2.6×10^{-4}
Rubber	10^{15}
P.V.C.	10^{15}

Example 7.1

Calculate the resistance of a 5 m length of copper cable of cross-sectional area 6 mm^2.

$R = \rho\ (l/a)$

therefore:

$$R = \frac{1.72 \times 10^{-4} \times 5}{6 \times 10^{-6}\ (\text{to convert } mm^2 \text{ to } m^2)}$$

$= 1.43\ \Omega$

However, the resistance of most materials changes with their temperature. Copper, for example, has an increase of 0.0043 Ω (for each ohm of resistance at 0°C) for each rise in temperature of 1°C, and this should be taken into account in calculations. This change is called the temperature coefficient of resistance (α) and the resistance (R) at a given temperature (T) can be calculated by

$R = R_0 + R_0\ \alpha\ T$

where R_0 is the resistance at 0°C.

Example 7.2

What are the respective values of resistance for the copper cable from example 7.1 at temperatures of 15°C and 25°C if the resistance at 0°C was 1.43 Ω?

(a) At 15°C: $R = 1.43 + (1.43 \times 0.0043 \times 15) = 1.52\ \Omega$

(b) At 25°C: $R = 1.43 + (1.43 \times 0.0043 \times 25) = 1.58\ \Omega$

Showing that the resistance to flow increases with a rise in temperature.

The electrical resistance of insulators, however, *decreases* with a rise in temperature (α is *negative* for these materials) and, in addition, the resistance of porous materials that absorb water (such as timber) *decreases* as the moisture content increases. This behaviour is used to good effect in the moisture meter, mentioned in chapter 3, which is based on electrical resistance. This can be used to assess moisture content not only in timber but also in other porous materials such as plaster and blockwork.

ELECTRICAL CIRCUITS

Electrical circuits consist of an energy supply (e.g. a battery), conductors (e.g. copper wire) and a load, which converts the electricity into some other form of energy such as heat, light, sound or movement. An electrical circuit (corresponding to the water circuit shown in fig. 7.1) is shown in fig. 7.2, together with some symbols

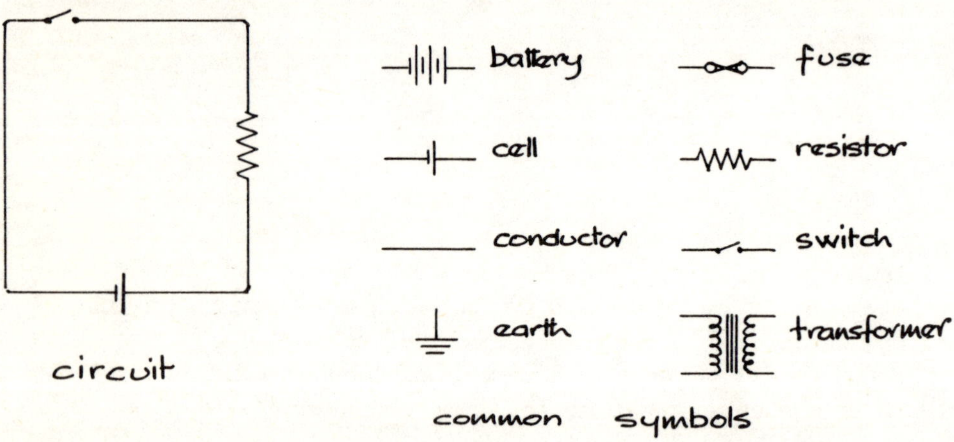

Fig 7.2 Simple electrical circuit and some symbols used in circuits

used in electrical circuitry. In this circuit the battery corresponds to the water pump, the switch to the tap, and the resistor (or load) to the paddle wheel. The conductor (or cable) will also offer some resistance in a similar fashion to the pipework. In both systems the flow occurs in one direction only. This is an example of a *direct current* (d.c.) [*Note:* Although, conventionally, current flows in the direction shown, the flow of electrons is opposite to this because they are negatively charged.]

Ohm's law

The relationship between current, potential difference and resistance is important and can be examined by using the circuit shown in fig. 7.3. For this experiment the resistor should be made of a material whose resistance change with temperature is negligible, since it may become warm during the operation. The current can be varied by using a variable resistance known as a *rheostat* (these are used in dimmer switches for lights, and for volume control in audio equipment) and values can be read from the ammeter and voltmeter, respectively.

By plotting a graph of potential difference against current, a straight line will be obtained (see fig. 7.4). The ratio of potential difference to current is therefore constant (being the slope of the line) and its value is the resistance R. Thus, for a resistance of this type, we can say $R = V/I$, or $I = V/R$, or $V = IR$. All these are statements of Ohm's law. From this we can see the following applications:

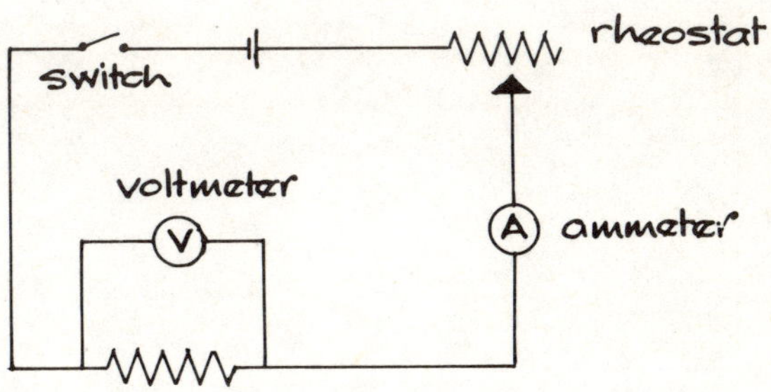

Fig 7.3 Circuit to verify Ohm's law

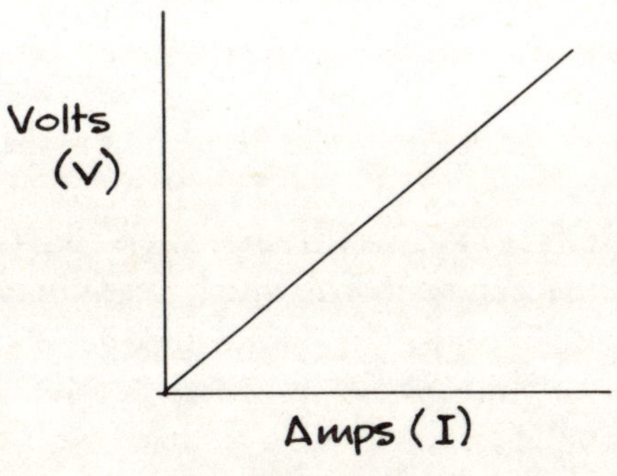

Fig 7.4 Relationship between current and potential difference

(a) If a voltage of 12 V is applied across a resistance of 10 Ω, what current will flow?

$$I = \frac{V}{R} = \frac{12}{10} = 1.2 \text{ A}$$

(b) What potential difference must be applied across a resistance of 5 Ω to cause a current of 6 A?

$V = IR = 6 \times 5 = 30$ V

However, Ohm's law is obeyed only if the resistance stays constant. Since there is a heating effect associated with an electric current (which will be discussed more fully later), the resistance, for many materials, will change as the current increases. An example of this is the filament of a lamp. Were the experiment to be repeated using a lamp, instead of the fixed resistance, a curve (rather than a straight line) would be obtained as shown in fig. 7.5.

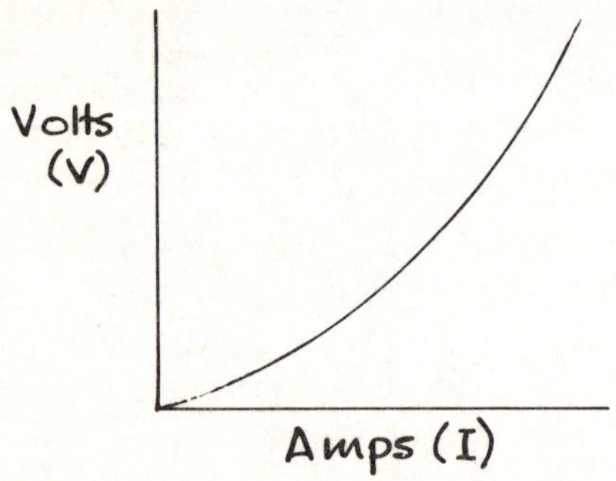

Fig 7.5 Current and potential difference for a filament lamp

Resistances in series

Let us consider what happens if there is more than one resistance in the circuit. When they are connected as in fig. 7.6, they are said to be in series, so that the same current flows through each resistance. By summating the potential differences across each resistance, we can obtain the total potential difference (in the same way that the total pressure drop across the tappings in the water circuit analogue equals the sum of the pressure drops).

When Ohm's law is applied to each resistance, then: $V_1 = IR_1$, $V_2 = IR_2$ and $V_3 = IR_3$ so that $V_{total} = V_1 + V_2 + V_3$, and $IR_{total} = IR_1 + IR_2 + IR_3$, so that, by cancelling the Is, $R_{total} = R_1 + R_2 + R_3$.

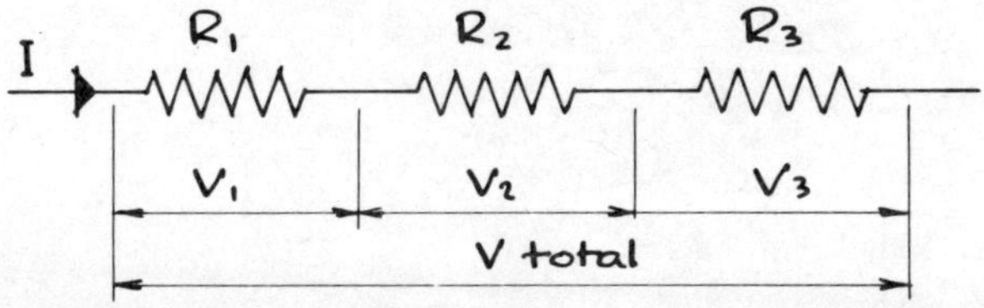

Fig 7.6 Resistances in series

This means that when resistances are in series they can be considered as a single resistance whose value is the sum of the individual resistances. For example, if resistances of 1 Ω, 2 Ω, and 5 Ω are connected in series, the total resistance will be $R = 1 + 2 + 5 = 8\ \Omega$.

Resistances in parallel

An alternative way of connecting resistances is shown in fig. 7.7, where they are said to be in parallel. Lighting circuits in cars, for example, are often connected in this way, so that in one lamp fails, current can still flow through the remainder. In this type of circuit, the potential difference across each resistance is the same but the current flowing in the rest of the circuit is split three ways at point A with the total current being the sum of the currents in the resistances. By applying Ohm's law again we see that:

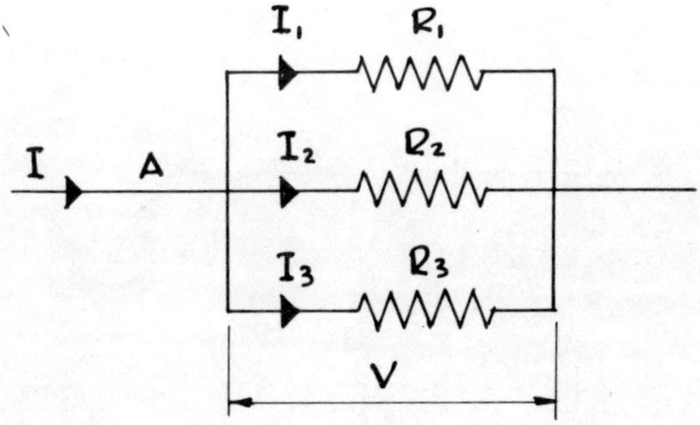

Fig 7.7 Resistances in parallel

$$I_1 = \frac{V}{R_1}, \quad I_2 = \frac{V}{R_2} \text{ and } I_3 = \frac{V}{R_3}$$

so that:

$$\frac{V}{R_{\text{total}}} = \frac{V}{R_1} + \frac{V}{R_2} + \frac{V}{R_3}$$

which gives:

$$\frac{1}{R_{\text{total}}} = \frac{1}{R_1} + \frac{1}{R_2} + \frac{1}{R_3}$$

Thus, if the resistances of the previous example were to be connected in parallel (rather than in series), the total resistance would be:

$$\frac{1}{R} = \frac{1}{1} + \frac{1}{2} + \frac{1}{5}$$

$$= \frac{10 + 5 + 2}{10}$$

$$= \frac{17}{10}$$

therefore:

$$R = \frac{10}{17}\,\Omega.$$

The wiring set-up in a domestic lighting circuit is a good example of a parallel circuit. The 'loop-in' method of wiring shown in fig. 7.8 ensures that equal pressure is available at all outlets and that failure of one bulb will not affect the rest of the circuit.

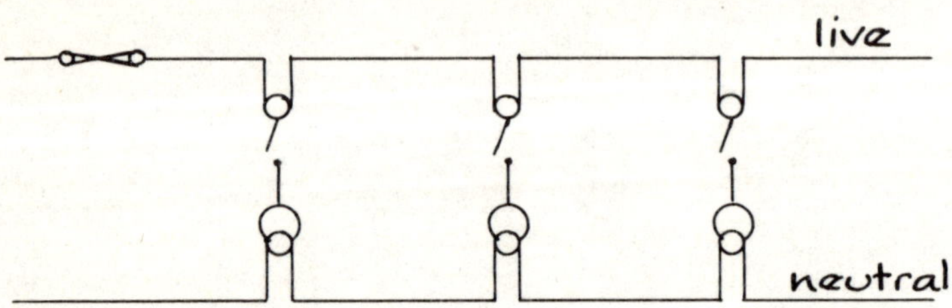

Fig 7.8 "Loop-in" method of wiring

ELECTRICAL POWER

Returning once more to the comparison with the water circuit in fig. 7.1, the rate of energy transfer to the wheel is governed by the head of water and the volume flow rate. Similarly the power consumed in an electrical circuit depends on electrical pressure and the rate of electron flow. The S.I. unit for power, as we know from chapter 4, is the watt, and the volt is defined such that one watt of power is consumed when a current of one amp is caused by a potential difference of one volt, so that watts = volts x amps. This is often translated into 'power = voltage x current' or $P = VI$. Useful alternatives to this can be obtained by substitutions involving Ohm's law, e.g. since $V = IR$, then $P = I^2R$ and since $I = V/R$, then $P = V^2/R$.

As explained in chapters 4 and 6, the kilowatt hour is a means of measuring power in cost terms. The amount of energy used by electrical appliances is invariably costed in this way.

Example 7.3

An electric fire has a rating of 2 kW and operates on a 240 V supply. What current does it take and how much will it cost to run at full output for 8 h (given cost of domestic electricity as 5.94p per kilowatt hour?

(a) Power = 2 kW or 2000 W. Since $P = VI$, $I = P/V$, so that current (I) = 2000 ÷ 240 = 8.33 A.

(b) Energy used = power x time = 2 kW x 8 h or 16 kWh. Cost = energy used x cost per unit rate = 16 x £0.0594 = £0.95 or 95p.

The benefit of knowing the cost is obvious but why do we need to know the amperage? We talked earlier about electricity causing heat. Any conductor carrying an electrical current generates heat. Sometimes this can be useful, as in the 'bars' of an electric fire or electrical elements buried in a concrete slab to provide underfloor heating, but often the heat is wasted. This is obviously undesirable from the point of view of energy conservation, as discussed in chapter 4, but it could also result in the overheating of cables, which is potentially dangerous. As a cautionary note, power or lighting cables in a domestic wiring system should not be covered by thermal insulation since this could raise the cable temperature to the extent where the electrical insulation material is melted and a 'shorting' of the circuit takes place resulting in the risk of an electrical fire!

Because the heating effect is proportional to the square of the current ($P = I^2R$), it is necessary to ensure that the safe carrying capacity of the circuit is not exceeded. This is equally true of the individual appliances at their connection to the circuit and these are 'fused' at the plug or spur socket. The fire in the example would need a fuse in excess of 8.33 A, which for practical purposes would mean using a 13 A fuse.

Protection of the circuits is carried out in two stages:

(a) The individual circuits, e.g. power to upper floor, power to electric cooker, power to ground-floor lighting, are protected at the 'consumer panel' by either fuses or circuit breakers.
(b) The main supply is protected by a large-capacity fuse on the supply side of the meter, its size being based on an estimate of the summation of appliances likely to operate concurrently.

A fuller description of a typical domestic installation is given in *Building and Civil Engineering Construction* Volume 2.

Another important aspect of the heating effect is the sizing of cable to be used in the circuit. As the current increases, so does the heat generated. In order to keep this to a safe level, the thickness of conductor needs to be increased. Typically, the cross-sectional areas used domestically are: 1.5 mm^2 for lighting; 2.5 mm^2 for power outlets; and 6.0 mm^2 for electric cooker.

Because of the dangers associated with electric shocks, it is important to protect the consumer from large currents. This is achieved by earthing the installations and appliances whereby, should a short circuit occur and exposed wiring become 'live', the earth wire provides the current with an easier path than that offered through the consumer. The low resistance of this path is often enough to cause a large current to flow which will blow the fuse.

CHEMICAL EFFECTS

We discussed these, in principle, in chapter 2 in relation to electrolytic action on metals. Also we looked at the conduction of electricity by solids earlier in the chapter. Solutions of acids, bases and salts are also conductors, not because they have free electrons but because they contain positive and negative ions (cations and anions, respectively), which can move when a potential difference is set up across the solution. Such solutions are called *electrolytes*.

A dilute solution of salt in water contains cations (H^+ and Na^+), and anions (OH^- and Cl^-). Suppose a circuit such as that shown in fig. 7.9 is set up. The positive electrode is called the anode and the negative one the cathode. The ions will migrate to the electrode of opposite charge, as shown. What happens at the electrodes depends on those used, the ions present and the concentration of the solution, but in the given example the following occur:

at the cathode $2H^+ + 2e \rightarrow H_2$

at the anode $2OH^- \rightarrow O_2 + H_2O + 2e$

The ions H^+ and OH^- are discharged (in preference to Na^+ and Cl^-) with the net effect that the water is split by electricity. The process is called *electrolysis* and many useful chemical changes can be brought about using similar methods.

Electrolysis is used in the refining of metals such as copper and silver where, for some electrical components, a high degree of purity

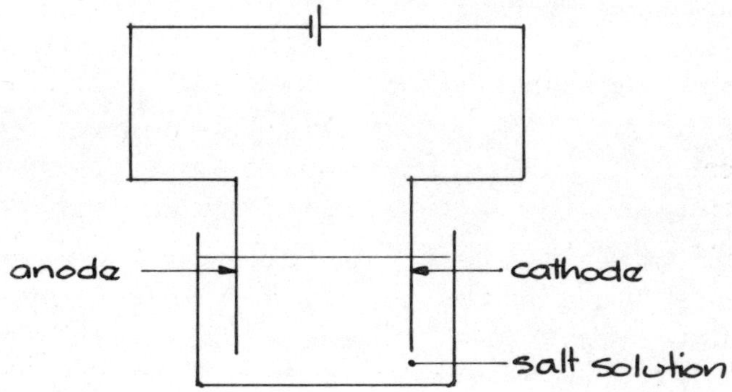

Fig 7.9 Electrolysis of a dilute salt solution

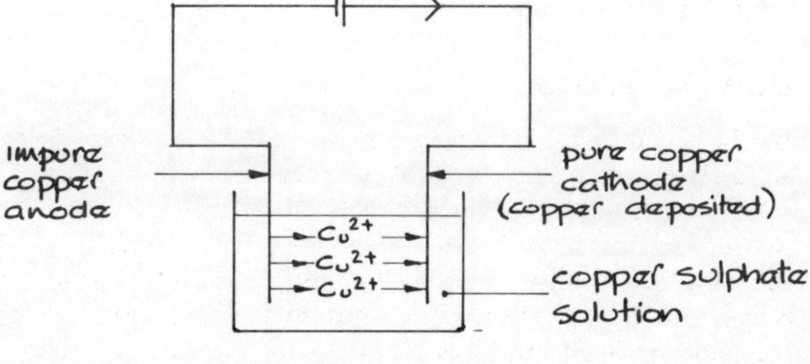

Fig 7.10 Electrolytic refining of copper

is required. In the case of copper the purification is carried out by passing a current through a copper sulphate solution, using copper electrodes as shown in fig. 7.10. In simple terms the electrode processes are:

anode $Cu \rightarrow CU^{2+} + 2e$

cathode $Cu^{2+} + 2e \rightarrow Cu$

The net result is the transfer of pure copper on to the cathode from the impure anode.

Another useful electrochemical process, mentioned briefly in chapter 2, is the formation of a thick protective film of oxide on aluminium. The component to be treated is made the anode of a 'cell', using dilute sulphuric acid as the electrolyte, and electrons are removed from the aluminium:

$Al \rightarrow Al^{3+} + 3e$

The aluminium ions then form the oxide film deposited on the surface of the metal. The process is known as *anodising* and is widely used for treating aluminium-framed doors and windows.

Cells

Primary cells

Just as electrical energy can be used to bring about chemical changes, so chemical changes can be used to produce electricity. A primary cell which consists of two different conductors is an electrolyte as shown in fig. 7.11. Since zinc is a more active metal than copper (see chapter 2, table 2.3), it has a greater tendency to lose electrons. These then flow round the outside circuit as shown.

For any pair of metals the polarity of the cell, and the size of the e.m.f. produced, depend on the positions of the metals in the electrochemical series (see table 7.2). The one higher in the series is the anode and the e.m.f. is the difference between the two electrode

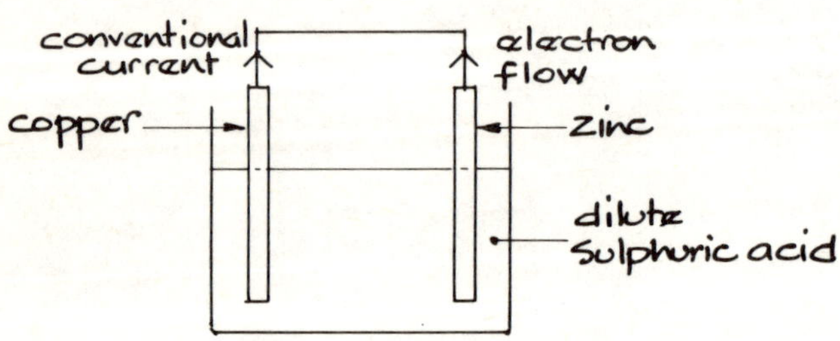

Fig 7.11 Primary cell

potentials. The cell shown in fig. 7.11 will have a potential difference (and thus an e.m.f.) of 0.76 + 0.34 or 1.10 V although this will depend on temperature and concentration of the electrolyte.

Table 7.2 Electrode potentials of metals (hydrogen as reference)

Element	Potential (V)
Sodium	–2.71
Aluminium	–1.67
Zinc	–0.67
Iron	–0.44
Lead	–0.13
Hydrogen	0
Copper	+0.34
Silver	+0.80
Gold	+1.68

One of the products of the cell shown is hydrogen gas, formed at the cathode. This tends to form a protective film around the metal and the e.m.f. will 'tail off'. This effect is known as *polarisation* and is undesirable in a practical cell. The dry cell (see fig. 7.12), as used for torch batteries, contains a depolarising substance (manganese dioxide) and so produces a fairly constant e.m.f. This is dependent on the cell being used intermittently to allow the depolariser sufficient time to work. A more compact cell, the mercury cell (see fig. 7.13), produces no polarising gases and thus gives a steady e.m.f., even in constant use.

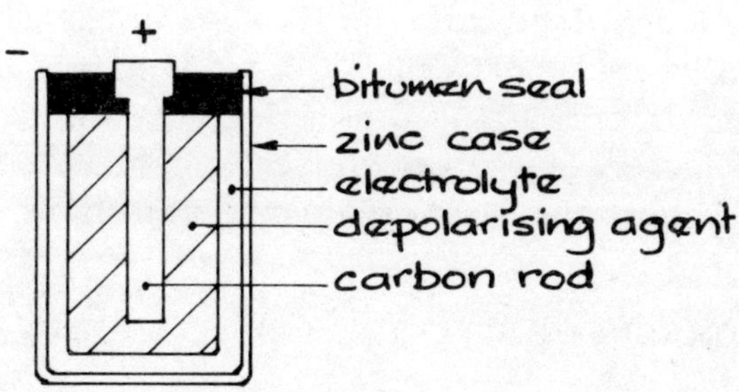

Fig 7.12 The dry cell

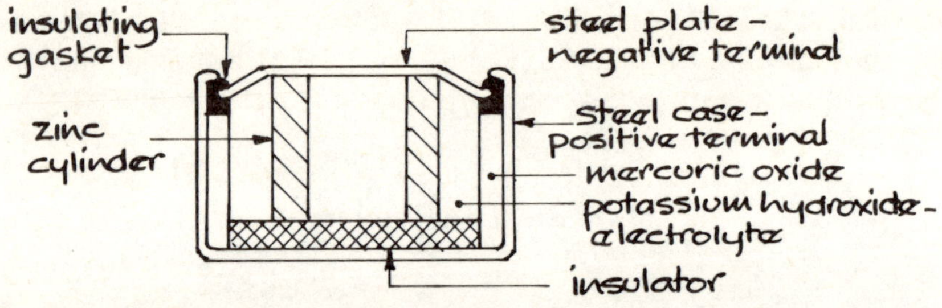

Fig 7.13 The mercury cell

Secondary cells

A secondary cell is a device for storing electrical energy and releasing it by chemical change. Unlike the primary cell, a secondary cell can be recharged by passing a current through it: this can be done many times during the life of the cell. In the fully charged state, the anode is lead dioxide (PbO_2) and the cathode lead (Pb). The electrolyte is dilute sulphuric acid (H_2SO_4). As discharge occurs, both plates are converted to lead sulphate ($PbSO_4$). The charge–discharge cycle of the lead–acid cell is as follows.

Fully charged		*Fully discharged*
Positive plate PbO_2	Discharge →	Positive plate $PbSO_4$
Negative plate Pb	← Charge	Negative plate $PbSO_4$
Electrolyte H_2SO_4		Electrolyte H_2O

The density of the electrolyte decreases as the cell is discharged and this gives a useful indication of the state of charge. Since the nominal voltage of a lead–acid cell is 2 V, this means that a 12 V battery comprises six cells arranged in series. An obvious application is that of a car battery but other applications include power for stand-by lighting systems and short-range communications.

ELECTROMAGNETIC EFFECTS

The relationship between electricity and magnetism has important applications, both in the way electricity can be generated and in its many uses. Any conductor carrying a current will have a magnetic field associated with it which can be plotted using a compass needle or iron filings (see fig. 7.14). The direction of the magnetic field can be found by imagining a screw being turned in the same direction as the current (see fig. 7.5). This effect is employed in the operation of lifting magnets, used in processing work, as shown in fig. 7.16.

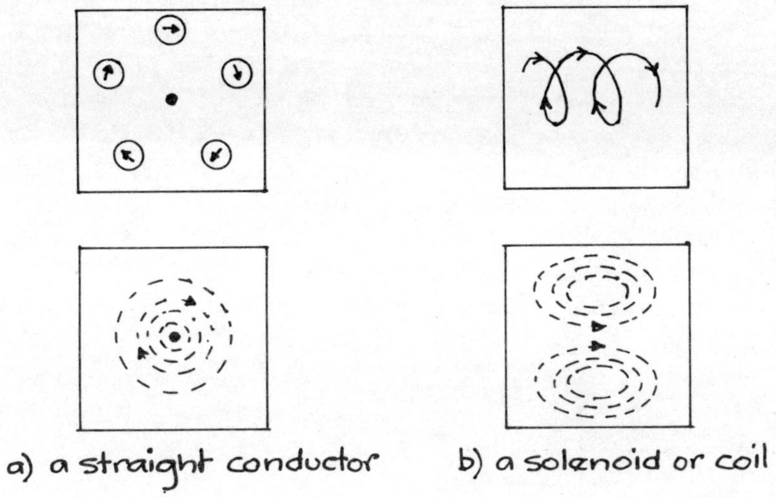

Fig 7.14 Magnetic fields round conductors

Fig 7.15 Direction of magnetic field

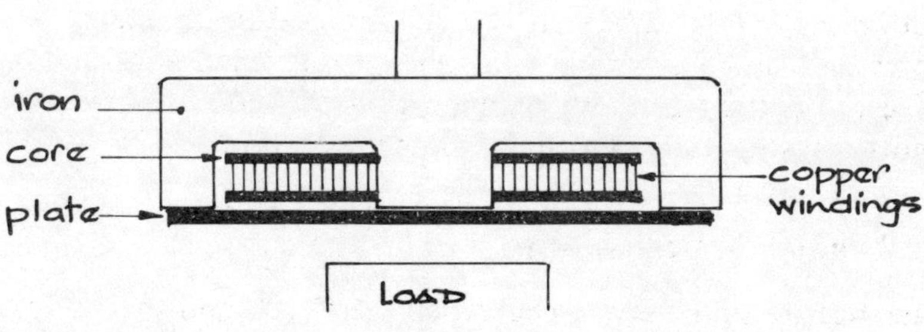

Fig 7.16 An electromagnet

Forces on conductors in magnetic fields

From the behaviour of the compass needles in fig. 7.14 we can deduce that there must be a force exerted on them. An equal and opposite force must, therefore, be exerted on the conductor, and this force can be utilised to cause motion. The relative directions of field, motion and current are shown in fig. 7.17. The moving coil meter (fig. 7.18) can be used to measure direct currents and voltages. When a current passes through the coil, there is a turning effect on it. As the coil turns, its movement is increasingly opposed by the spring so that, when a constant current flows, the turning effect of the coil will be balanced exactly by the torque of the spring. The magnitude of the deflection is directly proportional to the current and therefore the meter has a uniform scale.

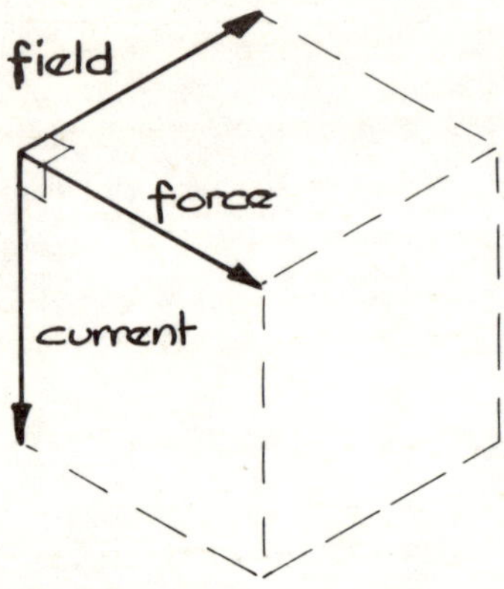

Fig 7.17 Directions of field, motion and current

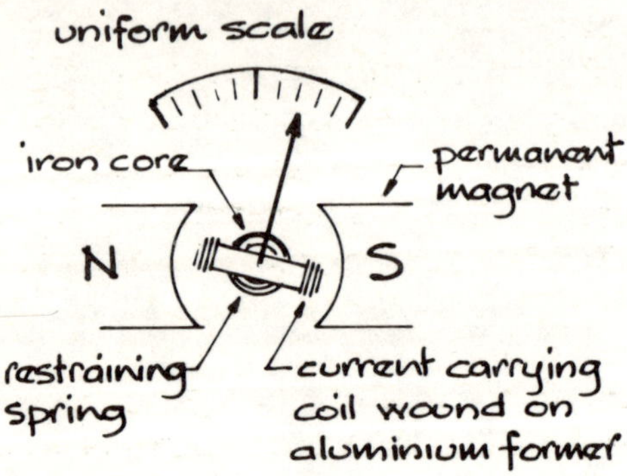

Fig 7.18 Moving coil meter

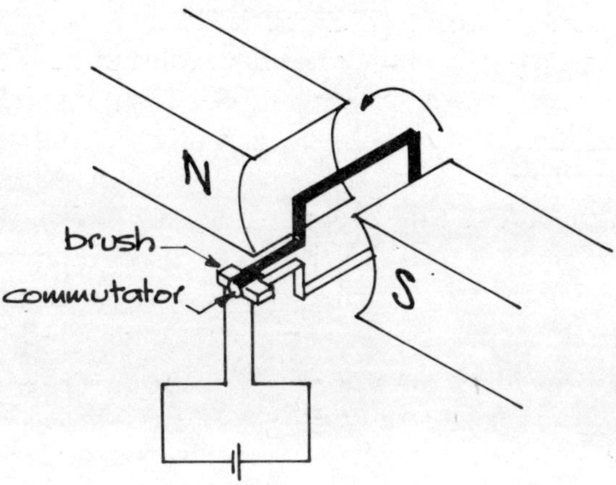

Fig 7.19 A simple direct current motor

Additionally, electric motors rely on motion caused by a current flowing in a magnetic field. A simple example of this is shown in fig. 7.19. The commutator reverses the current direction each half turn so that the coil continues to rotate in the same direction. More practical motors are discussed later in the chapter.

Electromagnet induction

Just as a magnetic field and a current can produce motion, so, by moving a conductor (forming part of a circuit) in a magnetic field, an e.m.f. can be induced in the conductor and a current will flow. This effect is called electromagnetic induction and we shall consider how electricity for domestic and industrial use can be produced in this way.

The direction of the induced e.m.f. in the conductor depends on its direction of motion in the magnetic field. This can be demonstrated by moving a magnet in and out of the coil (see fig. 7.20). The 'kick' on the meter is reversed each time the magnet changes direction; no deflection occurs when the magnet is stationary; and an e.m.f. is produced only when the conductor cuts across the magnetic field.

Generation of electricity

A simple way of producing electricity by induction is shown in fig. 7.21. This generator comprises a coil, made of copper wire, which is turned by an external power supply. The current produced can be 'tapped off' from the slip rings, which are in constant contact with

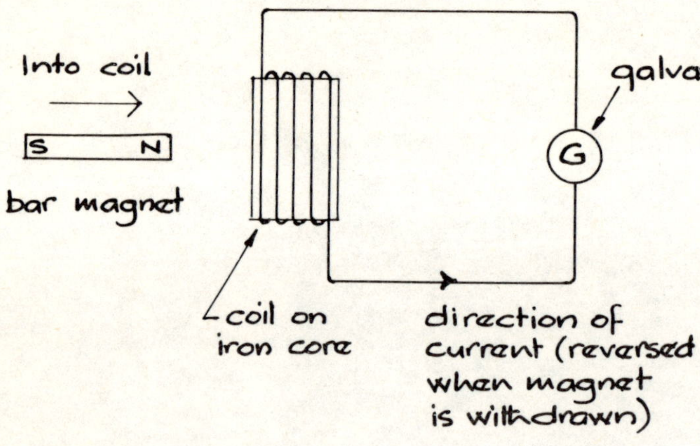

Fig 7.20 Electromagnetic induction

the coil. The size of e.m.f. is dependent on the rate at which the coil cuts across the magnetic field. As the coil turns through one revolution, the voltage produced will be that shown in fig. 7.22, with the direction of current being reversed at each half turn. This is called *alternating current* (a.c.), whereas the supply of current from the cells considered earlier flowed in *one* direction and was *direct current* (d.c.).

The number of voltage cycles per second depends on the speed of rotation of the coil. In the U.K. the mains electricity supply has a frequency of 50 Hz, which indicates that there are 50 cycles of volt-

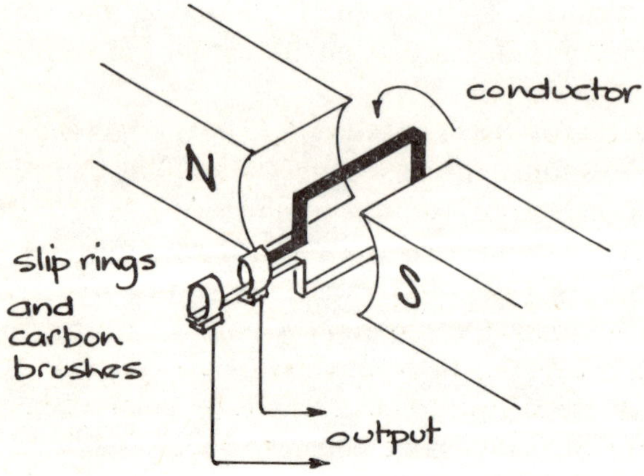

Fig 7.21 A simple generator

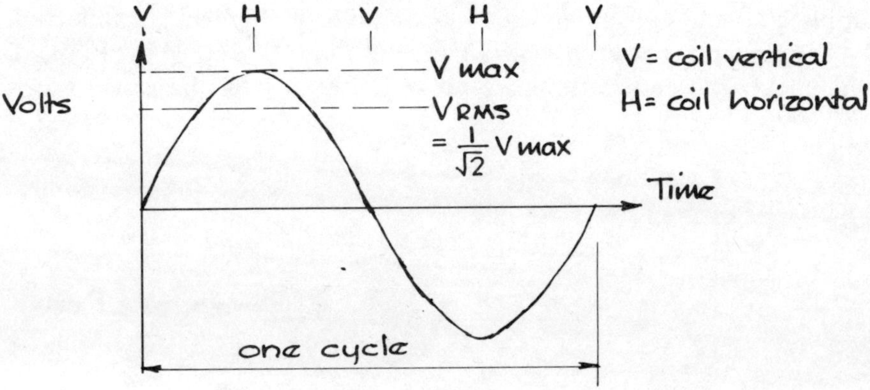

Fig 7.22 Output from a generator

age per second. The voltage produced by an a.c. supply is not constant, as can be seen from fig. 7.22, and, indeed, the average voltage over one cycle is zero. We have seen that electrical power depends on volts^2 ($P = V^2/R$) so the power consumption over a cycle will be that shown in fig. 7.23. It can be shown that the average power is half the 'peak' value, so the effective voltage producing this will be related to the square root of average power. This is called the *root mean square* (r.m.s.) voltage, thus, since

$$V_{r.m.s.}{}^2 = \frac{V_{max}{}^2}{2}$$

then

$$V_{r.m.s.} = \frac{V_{max}}{2}$$

This means that when we speak of a 'mains supply' of 240 V, we are talking of the r.m.s. voltage. The peak value of this is $240 \times 2^{1/2} = 339$ V. An a.c. supply of r.m.s. value 240 V has the same heating effect as that of a 240 V constant d.c. supply.

Practical generation of electricity

In practice it is easier to rotate the magnetic field, keeping the coil stationary, than to rotate the coil. The rotor of a generator is usually an electromagnet and three separate sets of coils form the stator. Since the coils are staggered (as shown in fig. 7.24), the three separate voltages produced will be out of phase. A common way of using a 'three-phase' supply is shown in fig. 7.25, where the value of voltage supplied to the consumer will depend on how the load is connected. By tapping off between one phase and neutral, 240 V is obtained, whereas, if the connections are made between two phases, 415 V is

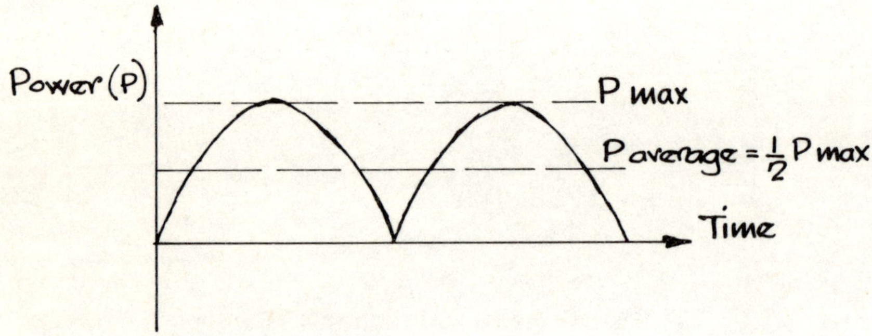

Fig 7.23 Power consumption by a resistance

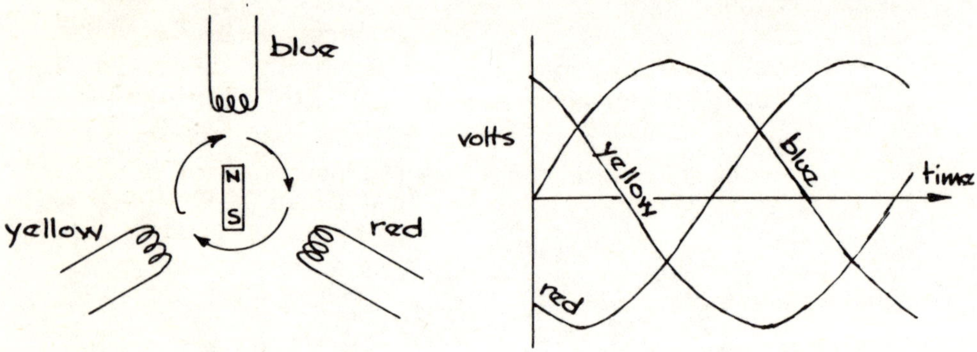

Fig 7.24 A "three phase" generator

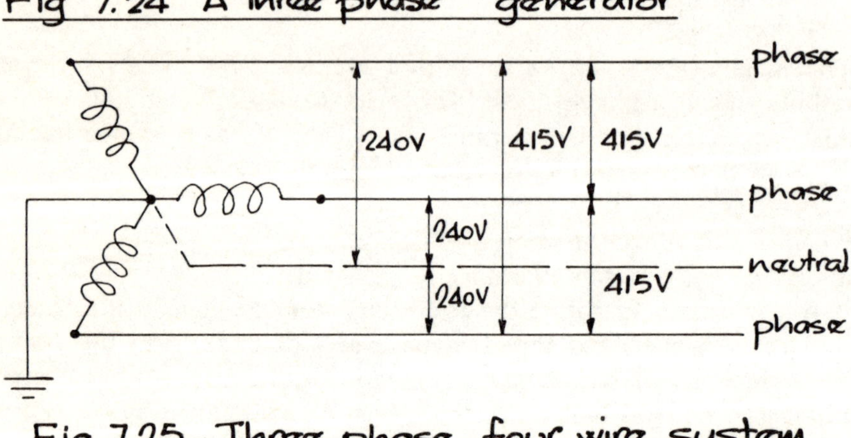

Fig 7.25 Three phase, four wire system

available. Domestic consumers require only one phase (240 V) but industrial consumers and building sites are likely to need three-phase supply, which is capable of delivering greater power. Also motors operating off a three-phase supply run more smoothly than those on single phase.

Electrical installations should be designed to ensure that all three phases carry an equal demand. When this is so, the current in the neutral wire is zero, allowing for economical selection of its cable size. This may, for example, be achieved by supplying every third house in a street with a particular phase in rotation.

Motors

The production of motion from a magnetic field and current has already been discussed but the type of supply available can be used in different ways. Some motors work off a d.c. supply, some off an a.c. supply, and universal motors work off either d.c. or a.c. Heavy motors on building sites will certainly need an a.c. supply but those used for portable hand tools are likely to be universal motors.

A common type of a.c. motor is the induction motor, which comprises a stator designed in such a way that the magnetic field produced by the current effectively rotates round the windings (remember how a rotating field can be used to produce a.c. electricity). The rotor may have a 'squirrel cage' winding with metal bars arranged in slots and connected at both ends to conducting rings. A typical example is shown in fig. 7.26.

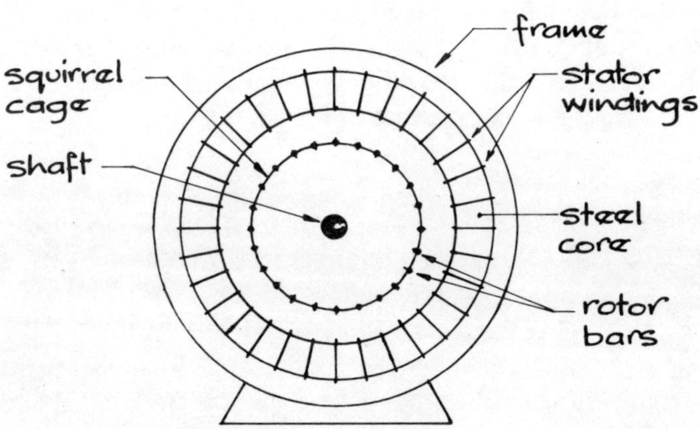

Fig 7.26 A "squirrel cage" motor

Transformers

A very important feature of a.c. electricity is that the alternating voltages may be increased or decreased ('stepped up' or 'stepped down') by transformers (d.c. cannot be so transformed). This is useful since it is desirable to transport electricity at very high voltages and then to step them down for use by consumers.

A simple transformer is illustrated in fig. 7.27 and consists of a soft

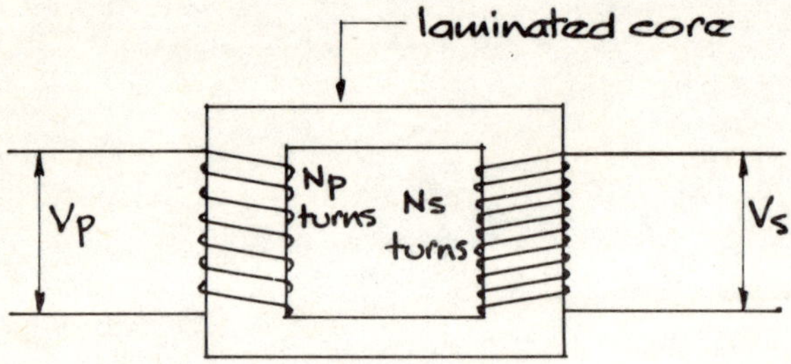

Fig 7.27 A simple transformer

iron core (laminated to minimise heat production) on which are wound two sets of coils, these being the input coil (primary) and the output coil (secondary). The alternating input voltage (V_p) induces a magnetic field in the iron, which in turn induces a voltage in the secondary windings. The voltage, per turn, is the same in both windings so that $V_p/N_p = V_s/N_s$. For a transformer with an efficiency of 100% (in practice some of the energy is lost in heat – see chapter 4), the power input is the same as the power output so that if the voltage is increased the current is decreased and vice versa, for example: $V_p \times I_p = V_s \times I_s$, and $V_p/V_s = I_s/I_p$.

Example 7.4

If a transformer steps down from 240 V to 100 V, and the primary windings have 1200 turns, how many turns are there in the secondary windings?

Since $V_p/N_p = V_s/N_s$, $N_s = (V_s \times N_p)/V_p$

$$= (110 \times 1200)/240 = 550 \text{ turns.}$$

The output voltage will depend on how it is 'tapped off' from the secondary windings since if it is 'centre tapped', the output will fall to 55 V. This is shown, together with the 110 V tapping, in fig. 7.28. An example of the application of a small transformer is the shaver socket in a bathroom, which provides the facility for both 240 V and 110 V appliances on separate sockets.

On building sites, where 240 V is considered unsafe for general use with small hand tools, the voltage is stepped down to 110 V but is additionally 'centre tapped' to earth to ensure that the maximum potential to earth is only 55 V, thus providing additional safety.

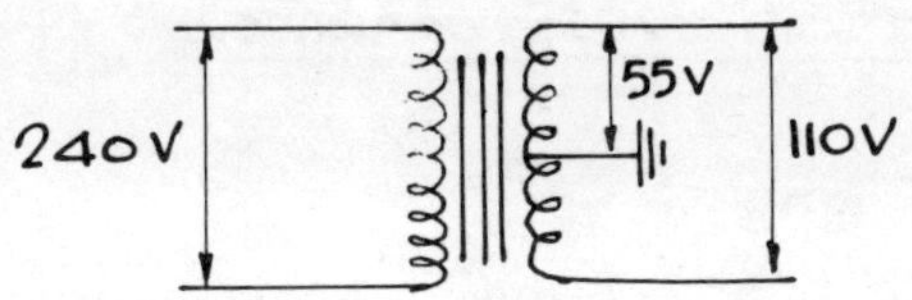

Fig 7.28 Tapping of transformers

GENERATION AND DISTRIBUTION OF ELECTRICITY

Regardless of whether coal, oil or nuclear fuel is used, all provide a means of converting water in boiler tubes into high-pressure steam, which can be used to drive the shaft of a turbine. This shaft turns a rotor, which is an electromagnet, magnetised by a small generator. Electricity is produced in the stator windings shown in fig. 7.29. We have already seen that heat produced in a conductor is proportional to I^2 so, if the current in the distribution system is kept small, less energy is wasted in transmission. Since power is the product of voltage and current ($P = VI$), it follows that a high voltage is required to provide a low current. For this reason, electricity is transmitted at high voltages. Typically, a power station generates at a pressure of 33 kV, which is then stepped up by transformers to 132 kV, 275 kV or 400 kV and fed into the grid system that links power stations. Area boards take off power from the grid at bulk supply points and distribute it over their areas at 33 kV. Further transformation to 11 kV occurs at primary substations (some industry takes its supply at this pressure), and to 240 V (415 V, three phase) at local substations (see fig. 7.30).

Site electricity

The construction site has many uses for electricity, and a 415 V three-phase supply is ideal for heavy plant, such as cranes, pumps, compressors and concrete mixers, whereas a 240 V single-phase

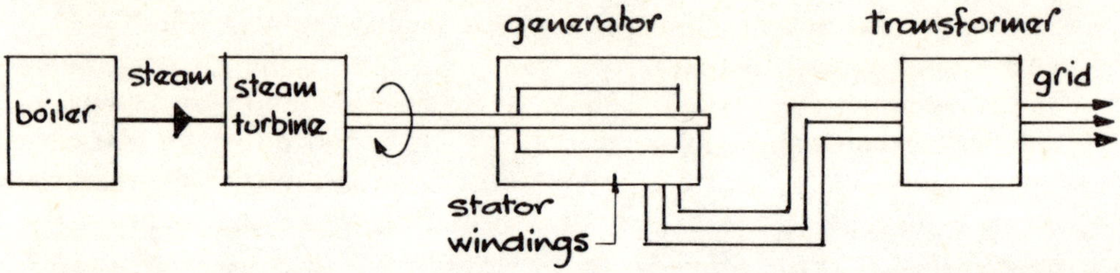

Fig 7.29 Generation and transmission of electricity

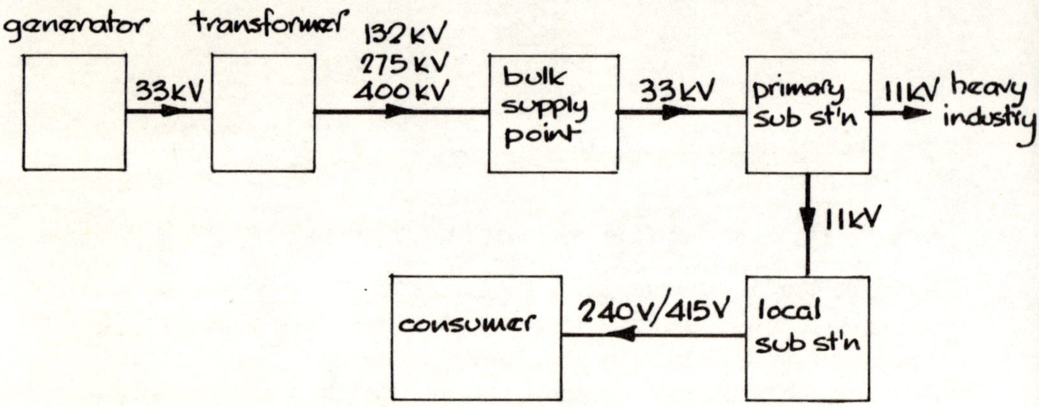

Fig 7.30 Transformation and distribution of electricity

supply is used for office accommodation and floodlighting. As mentioned earlier, a low-pressure supply of 110 V is preferred for local lighting, hand tools and 'crawler' lamps. This may be from either single phase or three phase via a step-down transformer.

Site distribution, ideally, will comprise the following items:

(a) an incoming supply unit, which incorporates the board's cable, fuses and meter;

(b) a main distribution unit supplying 415 V and 240 V;

(c) transformers to produce a 110 V supply from 415 V and/or 240 V;

(d) outlet units into which appropriate 110 V circuits are connected.

The maximum current in any of the above items must be adequately controlled by switches and either fuses or circuit breakers.

[*Note:* It is worth remembering that it is not *always* possible to obtain a temporary supply from the board and many sites rely on petrol- (or diesel-) driven generators to provide electricity. Also, not all countries are as fortunate as the U.K. in having a national grid and some outlying areas rely on wind-generated electricity. Other countries use their natural resources to good effect by using water from dams or waterfalls to create hydro-electric power stations. There is also some research into harnessing wave or tidal movements to generate electricity but it is important to consider the cost-effectiveness of such schemes.]

Exercises

7.1 Determine (a) the resistance of an appliance which draws a current of 5 A from a 240 V supply, and (b) the potential difference needed to cause a current of 2 A to flow through a 6 Ω resistor.

7.2 A potential difference of 12 V is applied to a circuit which includes resistance of 8 Ω, 4 Ω and 3 Ω.

(a) If the resistances are arranged in series calculate: the total resistance; the current in the circuit; the potential difference across each resistor.

(b) If the resistors are connected in parallel calculate: the total resistance; the current through each resistor.

7.3 Calculate the cross-sectional area of a copper wire 100 m long designed to carry a current of 5 A for a potential voltage drop of 4 V. The resistivity of copper is 1.72×10^{-4} Ωm.

7.4 What is the maximum power that can be drawn from a 240 V ring circuit that is protected by a 30 A fuse? What would be the hourly cost, working at this power, if electricity costs 5.94p per unit?

7.5 Given that the primary voltage of a transformer is 415 V with the number of primary and secondary turns as 1000 and 265 respectively, how may a voltage of 55 V be obtained from the transformer?

Assignment

Consider the case study 8.2 in *Building and Civil Engineering Construction* Volume 1. Describe the range of electrical equipment likely to be required during construction and, for each appliance, state the operating voltage and likely power consumption. Draw a diagram to indicate how the supply authority's electricity could be distributed on site and describe how this distribution system might be modified as the construction proceeds. Discuss how the end use of the buildings could determine the form of electricity distribution when the complex is completed.

8 Sound

Modern construction techniques have led generally to more lightweight buildings, particularly with respect to offices and, to an increasing degree, domestic properties. At the same time our environment has become much noisier, both externally and internally. Since these trends are likely to continue, more emphasis is certain to be placed on noise control than it has been in the past and it is clear that the construction industry has an important role to play in this regard.

The effect of 'noise', which can be defined as unwanted sound, range from the psychological (annoyance, fatigue and loss of concentration) to the physical (damage to hearing, structural damage from vibration).

The aim of this chapter is to investigate:

(a) the basic nature of sound;
(b) how sound is transmitted through materials;
(c) how sound levels can be measured;
(d) the details of construction that will afford some protection from both external noise and that produced in other areas within the same building;
(e) statutory requirements, relating to noise, that affect building construction.

Whereas light can travel through a vacuum, sound is transmitted only through a medium, which can be either solid, liquid or gaseous. This is easily shown in the experiment set up in fig. 8.1. As the air is slowly removed, the sound from the bell diminishes accordingly and any residual sound heard is that transmitted along the bell's supports.

THE NATURE OF SOUND

In order to understand more fully the properties of sound, it is

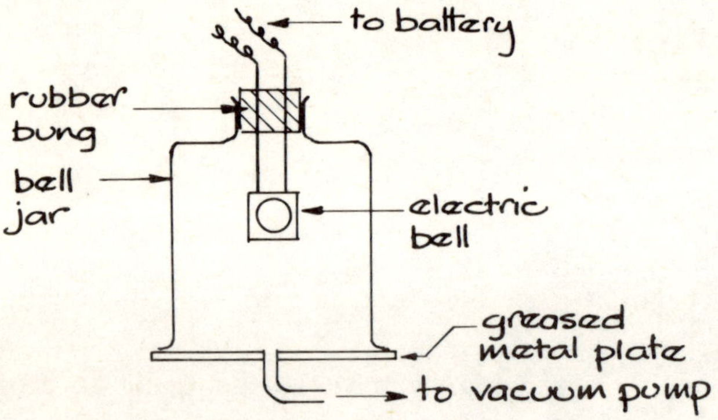

Fig 8.1 Sound transmission experiment

more instructive to consider the 'pure' sound produced when a tuning fork is struck. As the prongs move back and forth, the surrounding air is subjected to fluctuations in pressure, as shown in fig. 8.2. Because air is an 'elastic' medium, its pressure will tend to revert to its original value and when this happens the variations in pressure are passed on through the medium, producing a sound wave.

Any material that is capable of being stretched and compressed will behave similarly but, because air is so light, it is an excellent transmitter of such pressure waves and in any noise problem the 'air

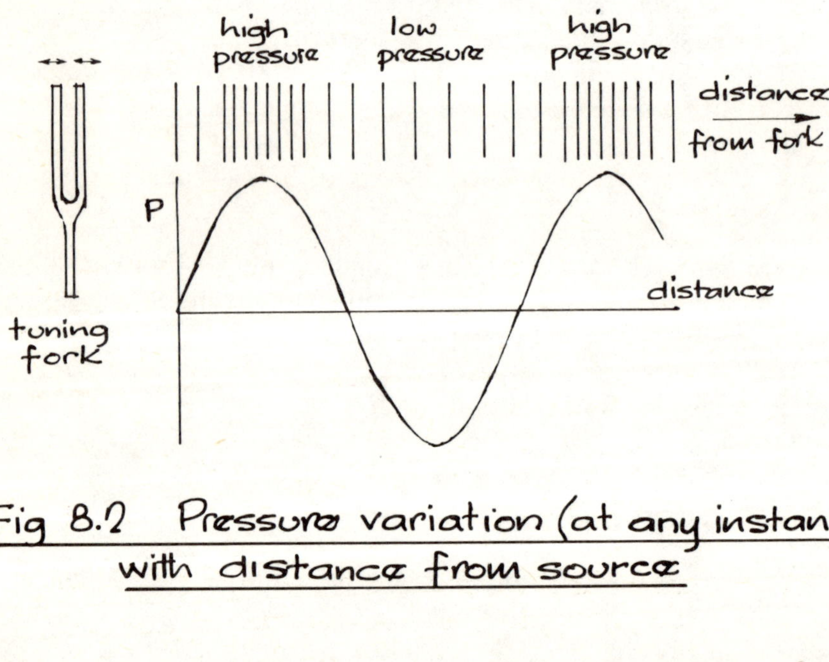

Fig 8.2 Pressure variation (at any instant) with distance from source

paths' are usually the most significant. The ear is sensitive to these changes of air pressure, which are transmitted via the ear drum, then through a series of bones to the inner ear where the signals are detected by nerve endings. These relay the signals to the brain, which recognises them as sound. It should be noted at this stage that the air is moving backwards and forwards in the same direction as the sound wave and this movement is described as *longitudinal wave motion*. If the displacement is at right angles to the direction of propagation (source), the wave motion is described as transverse. Electromagnetic radiation, such as light, is transmitted by this kind of wave. Both types are illustrated in fig. 8.3. The waveform shown in fig. 8.2 can be altered in two ways:

(a) If the tuning fork is struck more vigorously, the pressure variations in the surrounding air will be greater, as shown in fig. 8.4. The 'mean to peak' pressure change is known as the *amplitude* of the wave and, because the displacement of the eardrum will

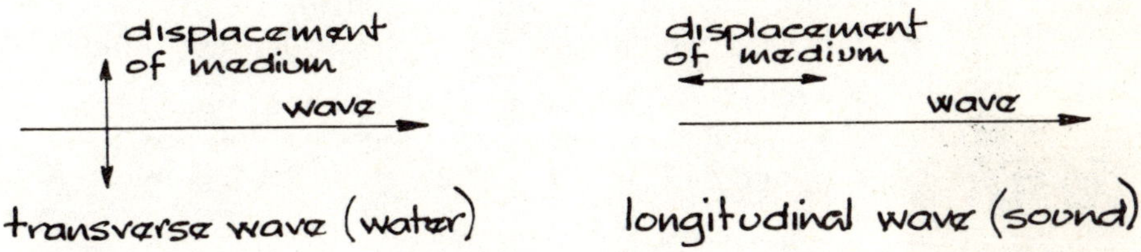

Fig 8.3 Wave motion

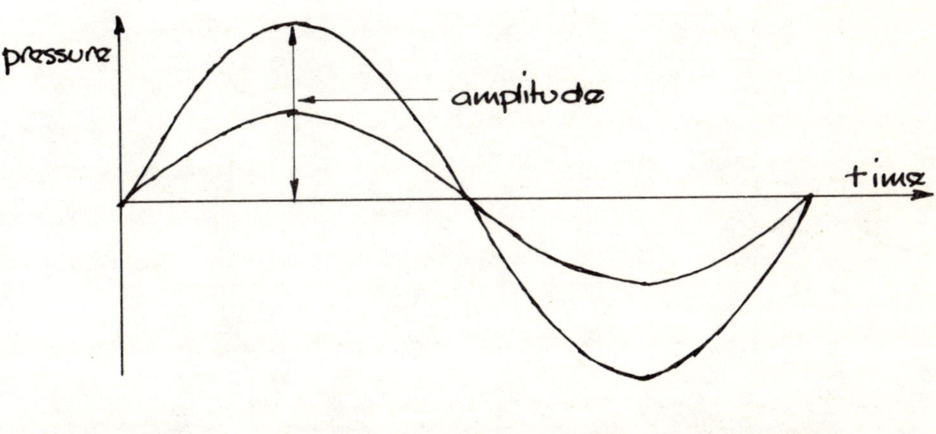

Fig 8.4 Amplitude related to loudness

increase accordingly, the change in amplitude will be perceived as an increase in the loudness of the sound.

(b) By using a different tuning fork we could change the rate of vibration of the prongs, which would change the number of pressure variations produced each second (see fig. 8.5). This quantity is known as the *frequency* of the sound and is measured in hertz (Hz). The frequency of the sound is related to 'pitch', which makes the sound appear lower or higher. [Note: You might notice that changing the frequency also changes the loudness of the sound. This is because the ear is not equally sensitive to all frequencies.]

Some frequencies of common sounds are given in table 8.1. The lowest and highest frequency figures are general and will vary depending on the age of the listener and any damage to hearing that he/she might have suffered. If we examine a section through the medium, in the direction in which the sound is travelling (as in fig. 8.6), the distance between adjacent pressure maxima (peaks) is called the wavelength of the sound and is measured in metres. In practice the wavelength can be the distance between any two identical points on adjacent cycles.

The velocity of sound is related to both frequency and wavelength. If a sound has a velocity of 330 m/s in air and a frequency of 200

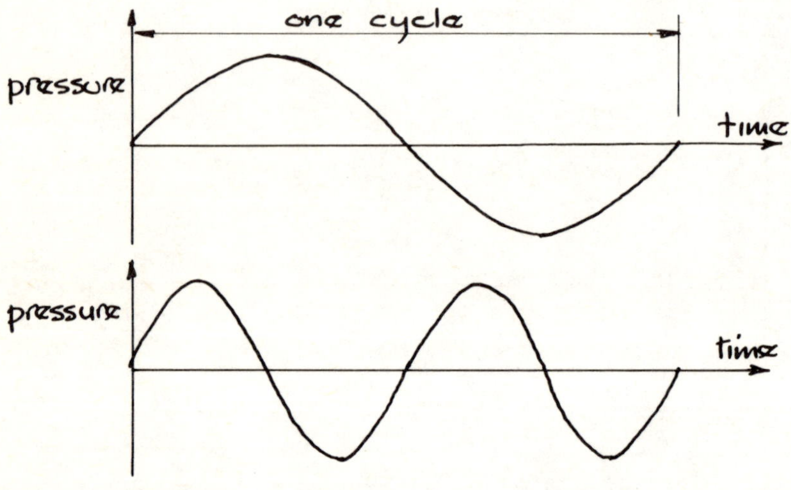

Fig 8.5 Frequency related to pitch

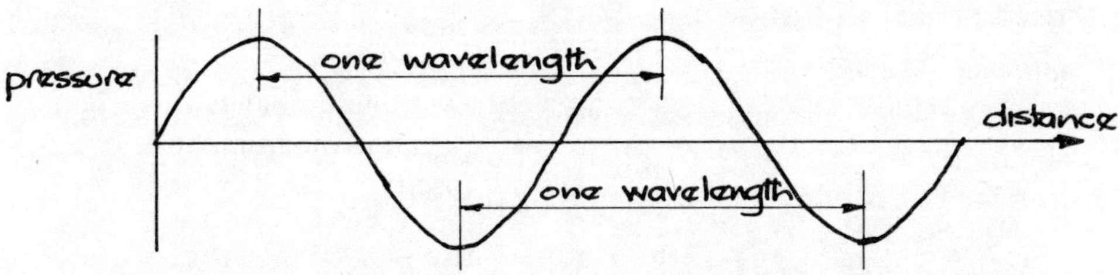

Fig 8.6 Measurement of wavelength

Table 8.1 Frequencies of common sounds

Frequency (Hz)	Nature of sound
20	Lowest frequency detectable by ear
50	Mains hum
262	Middle C
1 000	Radio time signal
20 000	Highest frequency detectable by ear
80 000	Sound emitted by bats

Hz, we can calculate that in 1 s there will be 200 pressure cycles and the sound will have travelled a distance of 330 m. Therefore the pressure maxima must be separated by a distance of 300 ÷ 200 = 1.65 m. This is illustrated in fig. 8.7. In general, therefore: velocity = frequency x wavelength, as mentioned in chapter 4.

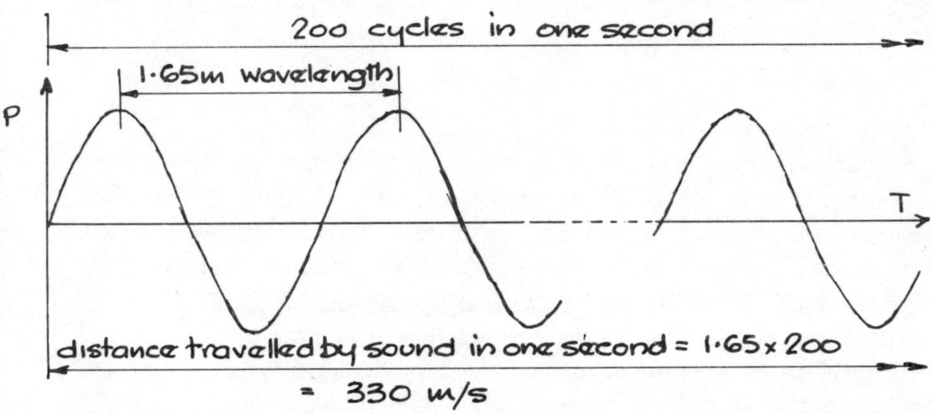

Fig 8.7 Relationship between velocity, frequency and wavelength

THE VELOCITY OF SOUND

The velocity of sound in air depends on both temperature and humidity. In other media the velocity is governed by how readily vibrations are transmitted and thus how easily the material may be stretched or compressed. Determining this will be:

(a) the density (ρ) of the material: the more dense the medium the more energy is needed to set the particles in motion;
(b) the modulus of elasticity (E). The stiffer the material, the more vibrations can be transmitted through it. It can be shown that: $V = (E/\rho)^{1/2}$.

As mentioned earlier (in chapters 2 and 4), this relationship is of value in the non-destructive testing of concrete, and the soundness of welds in metals, using an ultrasonic pulse velocity meter. The velocities of sound for some materials are given in table 8.2.

Table 8.2 Velocities of sound in common building materials

Material	Velocity (m/s)
Air	340
Low-density polythene	500
Lead	1100
Timber	3500
Concrete	3700
Steel	5200
Glass	5300

So far we have concentrated on sounds of a single frequency, sometimes called 'pure tones'. Most sounds, in fact, consist of a range of frequencies, and their waveform may be more complex than those discussed. Nevertheless, however complex the waveform appears, it can be made up by superimposing the more simple, pure tones. An example is illustrated in fig. 8.8. In order to analyse any noise problems, therefore, it is necessary to know the frequency content of the sound.

The problems of noise in buildings, however, are determined by noise levels, so we must establish how these may be described and

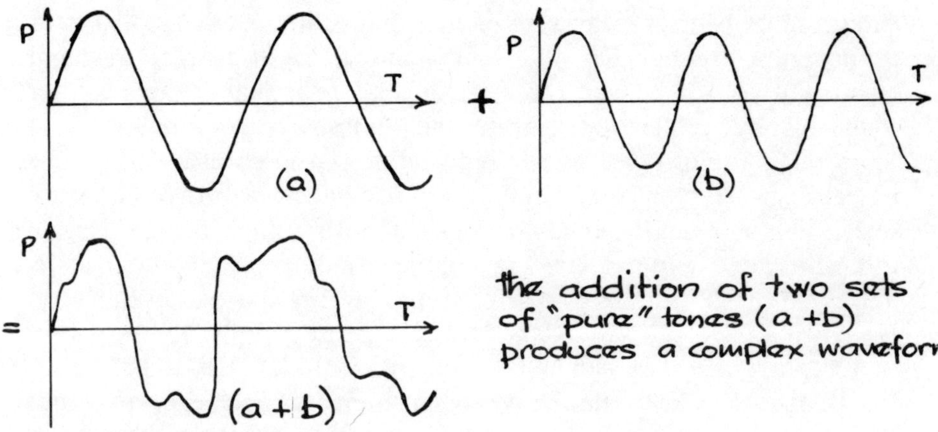

Fig 8.8 Production of complex waveforms

measured. We have seen how sound may be regarded as a pressure wave, so sound levels could be described using pressure units (newtons per square metre or pascals). Alternatively, if we remember that sound is a form of energy, then we could express the sound level by the rate of transfer of energy (watts) through unit area (square metres) of the medium. This is the intensity (*I*) of the sound and is therefore measured in watts per square metre. The energy stored in a sound wave is proportional to the area under the pressure curve, and so we can deduce that *I* is proportional to $\overline{p^2}$. Doubling the amplitude therefore leads to a fourfold increase in intensity. The relationship is illustrated in fig. 8.9.

The average human ear is capable of detecting sound pressure levels of about 20 micropascals (20×10^{-6} Pa) compared with a typical atmospheric pressure of 10^5 Pa. This level corresponds to an intensity of 10^{-12} W/m^2 and is called the *threshold of audibility* or

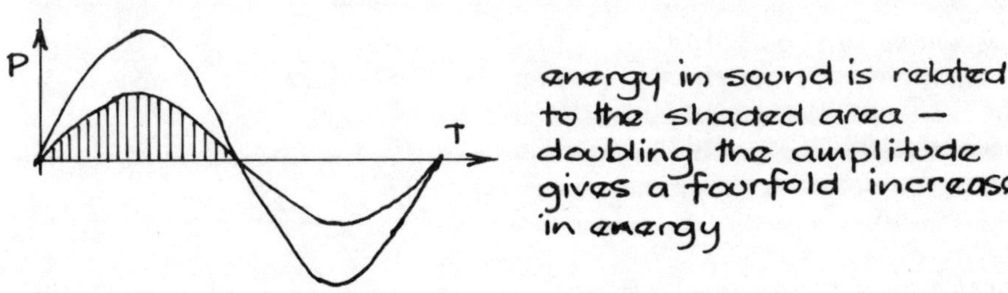

Fig 8.9 Relationship between amplitude and energy

threshold of hearing. (Because of the change in sensitivity of the ear to different frequencies, the values stated are those for a sound frequency of 1000 Hz.) The ear can tolerate pressure levels greater than this by a factor of about one million (corresponding to an increase in intensity of one million million) and so using either the pascal or watts per square metre would lead to a rather unwieldy scale. Let us examine what happens if, rather than using absolute intensities, we express levels as multiples of threshold intensities. Table 8.3 gives some examples of sound levels expressed in this way. By writing the multiples in standard form, and considering the power of ten only, the scale can be reduced from $10–10^{12}$ to 0–12. This is the basis of the bel scale. However, an increase of 1 bel corresponds to a tenfold increase in intensity. The ear can detect much smaller increases than this and it is more realistic to divide the bel into ten smaller units, each being one decibel (dB). Conveniently a change of 1 dB is just noticeable. To express a sound intensity I in decibels above the threshold intensity I_0 the mathematical process is therefore as follows:

(a) Write down the ratio I/I_0.
(b) Find the logarithm (to base 10) of I/I_0. This is the same as expressing the ratio as a power of ten, and writing down the power only.
(c) Multiply the logarithm by 10. Thus dB = $10 \log_{10} (I/I_0)$.

For example,
if $I = 6000 \times I_0$: $I/I_0 = 6000$. dB = $10 \log_{10} 6000 = 10 \times 3.778 =$ 37.78 dB

Table 8.3 Sound levels as multiples of threshold intensity

Sound	I (W/m^2)	I/I_0	$\log_{10} (I/I_0)$	dB above threshold
Threshold of hearing	10^{-12}	1	0	0
Speech	10^{-7}	10^5	5	50
Car	10^{-4}	10^8	8	80
Pneumatic drill	10^{-2}	10^{10}	10	100
Jet plane	10	10^{13}	13	130

[*Note:* In the examples given, we have compared sound with the threshold of hearing and the decibel is not a unit in the accepted sense. It is merely a convenient way of expressing a ratio.]

The addition of sound levels is an important process that we shall now consider. Since the scale we have developed is logarithmic, sound levels in decibels may not be added directly. For example, two sounds of 30 dB do not produce a total level of 60 dB. Instead, each sound level must be expressed as a multiple of threshold intensity. The resulting intensities may be added and the total then expressed in decibels as in the previous case. For example, what is the total decibel level if 30 dB and 40 dB are added?

If $dB = 10 \log_{10} (I/I_0)$, then:

$$I/I_0 = \text{antilog}_{10}\ (dB/10)$$
$$= 10^{dB/10}$$

thus:

$$I = I_0 \times 10^{dB/10}$$

In this example, $30\ dB = 10^3 \times I_0 = 1000\ I_0$

Similarly, $40\ dB = 10^4 \times I_0 = 10\,000\ I_0$

$$1000\ I_0 + 10\,000\ I_0 = 11\,000\ I_0$$
$$dB = 10 \log_{10} (I/I_0)$$
$$= 10 \log_{10} 11\,000$$
$$= 10 \times 4.04 = 40.4\ dB$$

As a further example let us add 55 dB and 57 dB.

$$55\ dB = 10^{5.5} \times I_0 = 316\,228\ I_0$$
$$57\ dB = 10^{5.7} \times I_0 = 501\,187\ I_0$$
$$\text{Total} = 817\,415\ I_0$$
$$dB = 10 \log_{10} 817\,415$$
$$= 10 \times 5.91$$
$$= 59.1\ dB$$

A rather quicker way of arriving at the total level is shown in fig. 8.10. Having calculated the difference in the levels to be added, the increase on the higher level can be read off. Thus, in our first example, the levels differ by 10 dB, which produces an increase of 1.4 dB on the higher level. This gives the same result of 40.4 dB obtained by the more rigorous method. The graph is of particular use when there are many levels to be added, as in the case of levels obtained from a frequency analysis of sound. The process is illustrated in table 8.4. For example, find the total noise level produced by the electric drill monitored in fig. 8.12.

THE SOUND LEVEL METER

Let us look at how sound levels may be measured. Since noise usually consists of sounds of many frequencies, and because the ear

Table 8.4 Addition of sound levels using graph (fig. 8.10)

Frequency	Level (dB)	Sub-total	Sub-total	Total (dB)
63	47	57.4	85	87.6
125	57			
250	64	85.0		
500	85			
1000	82	83.5	84.1	
2000	78			
4000	75	75.0		

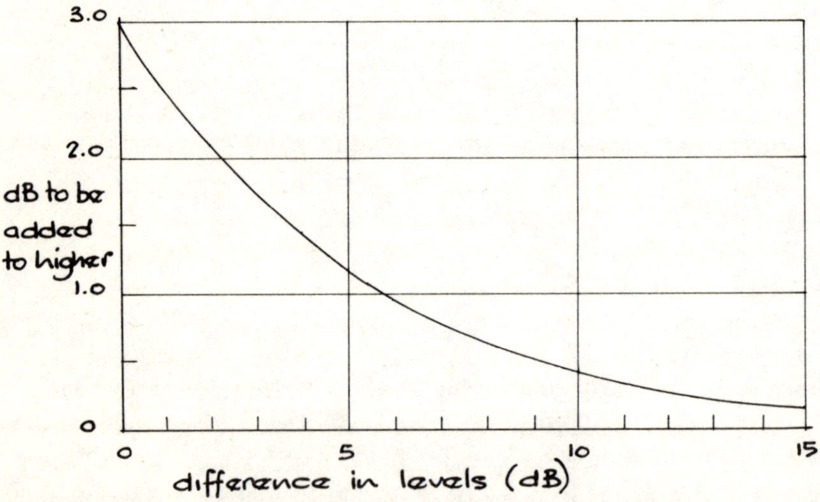

Fig 8.10 Addition of sound levels by graph

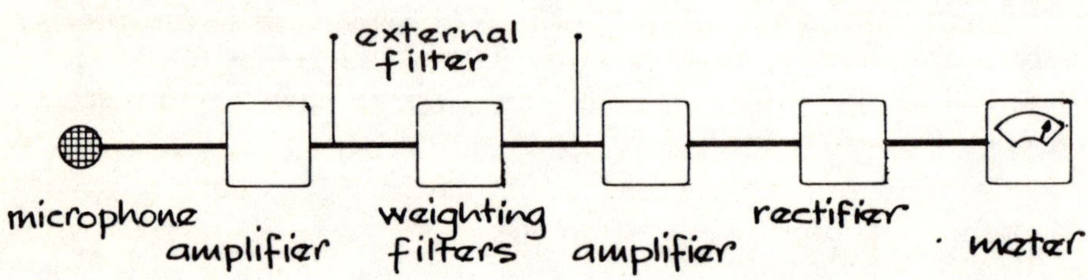

Fig 8.11 Diagrammatic sound level meter

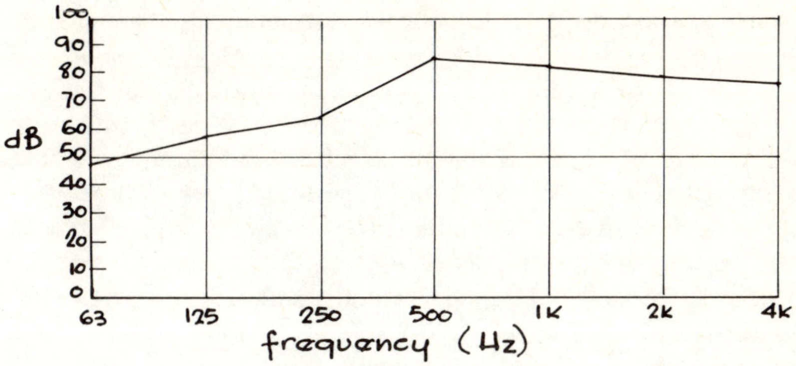

Fig 8.12 Octave band analysis

is not equally sensitive to all frequencies, a measurement that correlates well with the performance of the ear would be useful. The sound level meter shown in block diagram form in fig. 8.11 is capable of such measurement. The diaphragm of the microphone vibrates in response to the sound and, in doing so, creates a small electrical current whose size is related to the amplitude of the sound. The purpose of the amplifier is to make the current large enough to drive the display needle.

In many circumstances it is useful to be able to measure sound pressure levels over a range of frequencies and most meters will have either an inbuilt filter or connections so that an external filter can be used. In either case, the levels, in octave bands (or for more detailed analysis, third-octave bands), can be measured. The centres of the octave bands are the following preferred frequencies: 16Hz, 31.5 Hz, 63 Hz, 125 Hz, 250 Hz, 500 Hz, 1000 Hz, 2000 Hz, 4000 Hz and 8000 Hz. It should be noted that an octave interval represents a doubling of frequency. Also the widths of the bands are proportional to the mid-frequencies, so that they are not all of equal width. Thus, for example, if sound levels in the 125 Hz octave band are measured, all sounds, other than those in a narrow band centred on 125 Hz, are attenuated so that (in effect) they are ignored by the meter. A typical octave band analysis of a noise is shown in fig. 8.12.

The purpose of the 'weighting filters' is to make the meter respond in a similar fashion to the ear to sounds of different frequencies. By giving more 'weight' to frequencies at which the ear is more sensitive, a good correlation between the meter readings and subjective assessment of the loudness of the noise can be achieved.

Since the change in response with frequency also depends on the loudness of the sound, there are three weightings designated as: A, for low levels of loudness; B, for medium levels of loudness; and C, for high levels of loudness. For the majority of cases, such as traffic and industrial noise, the A weighting is used and sound levels obtained are expressed in dBA. When the weighting filters are not used, the meter responds linearly to sound pressure and it is used in this way when absolute measurements of sound pressure levels are needed, as in the case of determining the sound insulation properties of partitions.

SOUND INSULATION

Having now discussed the generation, propagation and measurement of sound, let us examine its behaviour in relation to construction. For the purposes of our discussion we shall define any element, such as a wall, floor or ceiling, as a 'partition'. What may happen if sound energy is incident on a partition, as shown in fig. 8.13? The relation-

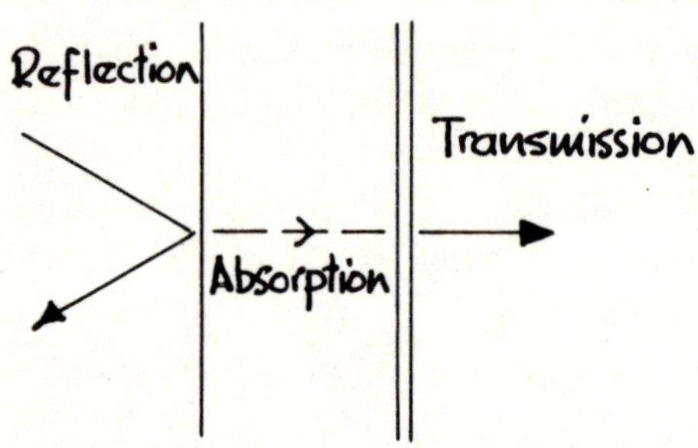

Fig 8.13 Sound incident on a partition

ship between incident and reflected sound is determined by sound absorption and is therefore of interest when the source of sound and the listener are in the same room. This aspect of sound will be developed in Volume 2. Here we shall investigate the factors influencing the reduction of sound levels across a partition (from one room to another, or through an external wall, for example). This is what is meant by sound insulation. In fact there is little correlation between the effectiveness of a partition as a sound insulator and its performance as an absorber of sound. Sound absorption tends to be influenced greatly by surface finishes and there are many ways of modifying the amount of *sound absorption* in a room after the building has been constructed.

On the other hand, *sound insulation* is more likely to be influenced by the materials and construction methods so that the control of sound insulation is mainly in the hands of the designer and contractor. For a sound to be heard through a partition, the partition itself must, in some way, have been set in motion. This may have happened by direct impact, such as a footstep, hammering, or dropping a heavy object, and is called *structure-borne sound*. Alternatively, the air on one side of the partition may have been set in motion by the sound from, say, audio equipment or a conversation. These vibrations are passed on to the partition and thereby transmitted to the air on the other side. In this case it is called *airborne sound*.

The distinction between structure-borne and airborne sound is an important one since the methods used to control them may be quite different. Since sound levels in decibels above threshold are comparative, we may use a similar approach to describe the airborne sound insulation of building elements. As an example, if the partition shown in fig. 8.14 has incident and transmitted levels of 80 dB and

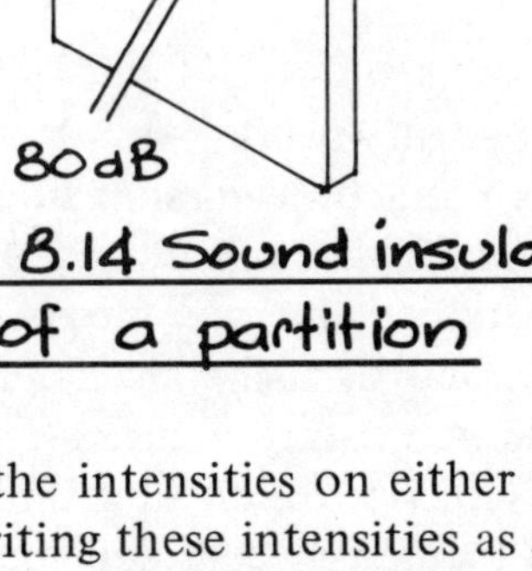

Fig 8.14 Sound insulation of a partition

30 dB, respectively, the intensities on either side will be give by $10^8 \times I_0$ and $10^3 \times I_0$. Writing these intensities as a ratio gives:

$$\frac{\text{Incident intensity}}{\text{Transmitted intensity}} = \frac{10^8 \times I_0}{10^3 \times I_0} = 10^5.$$

Clearly this factor of 10^5 corresponds to a change of 50 dB, but we

could have arrived at this result more simply by just subtracting the transmitted level from the incident level. In general, then, if the incident and transmitted levels are L_1 and L_2, the sound insulation of the partition will be given by $L_1 - L_2$.

It is worth noting that the value obtained will depend on the partition only, and the same reduction will be obtained no matter what the incident level L_1. Precise measurements of airborne insulation must be obtained with great care because results are so dependent on the methods used. A more detailed description will be given in Volume 2. Values of impact level, given in standards, refer not to a difference in levels but to the transmitted level only, so that a high level of transmitted sound indicates low impact insulation.

Figure 8.15 indicates some of the paths that sound can take in buildings and from the diagram we can see that not only does sound

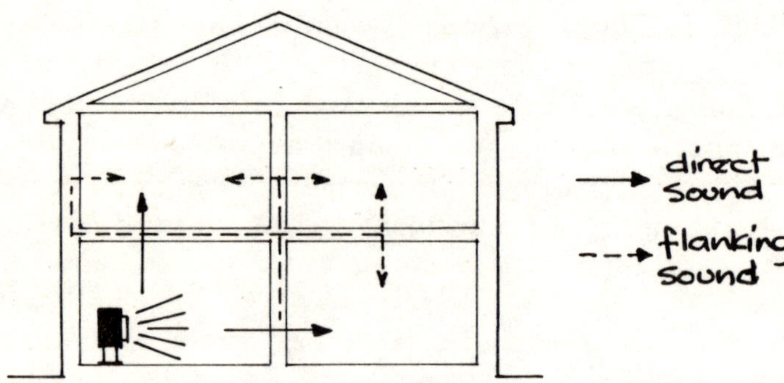

Fig 8.15 Sound paths in buildings

travel directly through walls and floors, but that it is capable of being transmitted into neighbouring rooms via a number of indirect paths. This is often called flanking transmission and it will determine the overall sound performance of the building. It is important to distinguish between laboratory measurements of sound insulation, where flanking is minimised by using specially mounted test panels, and field measurements in existing buildings, where the size and fixing of the partition is likely to differ from that in a laboratory. Attention to detail and quality control in construction can, therefore, improve the insulation significantly.

The airborne sound insulation of a partition will depend on a number of factors and these are now considered.

Mass

In order to vibrate a partition backwards and forwards, a force must be applied to it. For the type of vibration we are considering, its amplitude is proportional to the acceleration of the partition. The effect is shown in fig. 8;16. Applying Newton's second law of motion

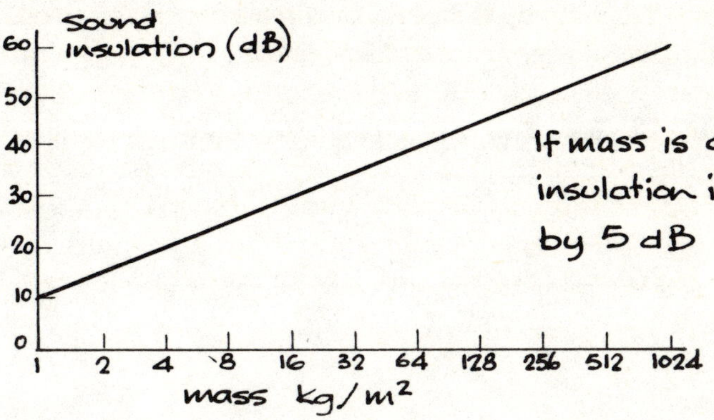

Fig 8.16 The mass law of sound insulation

gives: force = mass x acceleration. Therefore force $\propto$ mass x amplitude. Thus, for a given force (corresponding to the incident sound pressure), the amplitude of vibration is inversely proportional to the mass. This means that if the mass is doubled, the amplitude is halved. We have already seen that this would give a fourfold drop in transmitted intensity and consequently there will be an improvement in insulation of $10 \log_{10} 4$ dB = 6 dB. In practice, other forces are also at work and the actual improvement is 5 dB.

In order that different forms of construction may be compared for sound insulation, the 'mass per unit area' (or superficial mass) is considered, and if a graph of sound insulation against superficial mass is plotted for a large number of different types of partition, it is found that the line shown in fig. 8.16 is obtained. [*Note*: The mass scale is *not* linear.] This is a graphical representation of the mass law of sound insulation and shows that *if the mass per unit area of the partition is doubled, the insulation increases by 5 dB.* The values of sound insulation used in the graph are those averaged over a range of frequencies. It can also be shown that a similar increase in insulation is obtained if the frequency of the sound is doubled, and airborne insulation generally increases with frequency. If music is heard through a wall, for example, the bass notes are transmitted far more effectively than ones of higher pitch. Because of this variation with frequency, a detailed analysis of sound insulation performance would mean measurement using the 'third-octave bands' mentioned previously. As an example of the mass law, consider the insulation of the brick walls (plastered on both sides to a depth of 13 mm) given in table 8.5. In general, the mass of a partition is the most important factor in determining its sound insulation.

Table 8.5 Sound insulation values for solid brick walls

Thickness (mm)	Mass/unit area (kg/m^2)	Average insulation (dB)
112.5	270	45
225	480	50
450	920	55

Uniformity

If the brick wall of thickness 112.5 mm is *not* plastered, reducing the mass to 220 kg/m^2, its insulation is reduced to between 35 dB and 40 dB. This is a far greater drop than can be accounted for by mass alone and the chief reason for the drop in performance is the porosity of the brickwork and, possibly, gaps at the mortar joints. By sealing these air paths, the airborne insulation improves remarkably. Since air is such a good transmitter of sound, being light and elastic, even relatively small air gaps in partitions will seriously impair the sound insulation. As an example, an external door of solid hardwood panelled construction, 50 mm thick, has an insulation value of 25 dB when hung in a frame without draught stripping. If magnetic draught seals are applied to close the air gaps, the insulation increases to 30 dB.

More generally, if any partition contains an area of relatively poor insulation, its overall performance will be adversely affected since the sound paths will 'seek out' the easiest route. This will always be the case for doors and windows in walls and, where sound insulation is of prime importance, attention must be paid to these weak areas. Some obvious situations are offices located on busy traffic routes, housing near airports or motorways, and housing and offices near schools.

Because the weakness is related directly to air paths at the perimeters of the frames, it is important to seal, as effectively as possible, the gaps both at opening sections and at the junction of wall and frame. The latter can be achieved by mastic sealant, which also forms a weather seal, whereas the former is probably best dealt with by the manufacturer incorporating a Neoprene weather strip in the opening sections of the frame. These also eliminate draughts and allow better control of air changes, and so raise the comfort level generally. Alternatively, if the frames are already installed without such weathering, it is possible to add nylon or magnetic draught stripping as a remedial measure, which affords similar benefits.

Stiffness

At certain frequencies, single-leaf partitions, such as single-glazed windows, may show a marked dip in insulation. This happens when the partition vibrates along its length rather than acting as a membrane in the usual way. This effect is called 'coincidence' and is illustrated in fig. 8.17. It is governed largely by the stiffness-to-weight ratio of the material used and so the effect is not significant for heavy

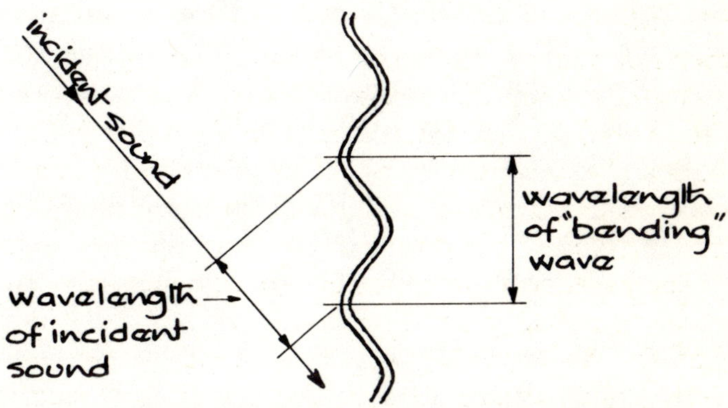

Fig 8.17 The coincidence effect

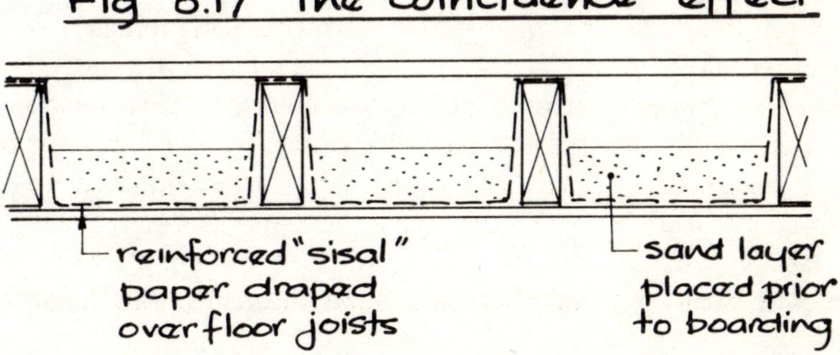

Fig 8.18 Wood joist floor with "pugging"

partitions. It is possible to increase the mass of a timber floor, for example by using 'pugging' (fig. 8.18), without increasing its stiffness, and in such cases the frequency at which coincidence occurs is raised to a value above the range that is of concern to the human ear. It should be realised that the stiffness is a property of the partition itself and is not necessarily related to its method of fixing. In general, the more rigidly the partition is fixed, the less readily will it vibrate.

Discontinous construction

It is possible to isolate parts of a structure so that vibrations are not easily passed between the component parts, and this may improve the overall performance with respect to sound insulation. A simple example of this is the vibrating window. If a Neoprene strip separates the opening light from the frame, it will 'damp' the vibration between the two components in addition to its other functions. This means that it contributes to both thermal and sound insulation in a variety of ways.

Two commonly used forms of discontinuous construction are cavity walling and double glazing. If a cavity wall, comprising two leaves (each with an insulation value of 45 dB), could be constructed to provide complete structural isolation, the overall insulation would be 90 dB, as shown in fig. 8.19. In practice this cannot be achieved since the two leaves are connected at frequent intervals by wall ties, which means that the insulation value is more likely to be about 53 dB (which is not vastly different from that of a solid 'one-brick' wall). Also the other functions of the wall must be considered, such

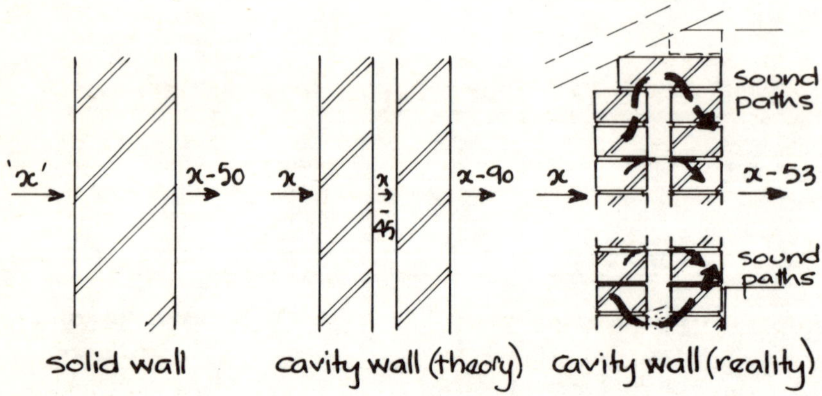

Fig 8.19 Sound insulation - brick walls

as thermal insulation, which means that the inner leaf is likely to be built in lightweight aerated concrete rather than brickwork. The cavity does, however, provide a small contribution to sound insulation.

Double glazing, however, does show a dramatic improvement in sound insulation, as can be seen in table 8.6, and this increases with a corresponding increase in air gap. If double glazing is out of the question, some improvement can be effected by using heavier glass (possibly 'plate' glass) but high levels of insulation can be achieved with standard 4 mm thick glass when an air gap can be provided. Unlike the cavity wall, there can be complete discontinuity between the two panes, and if they are in separate frames, the air space can be made even more effective. Thermal insulation double glazing has an optimum gap of about 20 mm, but table 8.6 illustrates that for sound insulation, the wider the gap the better. This is of particular interest in office buildings where the overriding requirement is for sound rather than thermal insulation (heating costs can be passed on to the customer but the products of physical discomfort are best *not* passed on).

Table 8.6 Sound insulation properties for various windows

Type of window	Insulation (dB)
Single glazed	
4 mm glass	24
6 mm plate glass	27
12 mm plate glass	30
25 mm plate glass	35
Sealed unit double glazing (4 mm glass)	
12 mm air gap	32
20 mm air gap	34
Double casements (4 mm glass)	
50 mm air gap	36
100 mm air gap	39
200 mm air gap	43
300 mm air gap	46
400 mm air gap	49

As with single glazing, all of this is of little value if the detail at the edges of the windows allows air passages to carry noise. The perimeters should, therefore, incorporate draught sealing, 'damping out' of vibration and sound-absorbent reveals.

For domestic construction there is very little difference between sealed-unit double glazing (20 mm air gap) and secondary glazing (50 mm air gap) and for new work the convenience in use of the former outweighs the 2 dB improvement. Where noise is a real problem, however, such as near an airport or motorway, it might be sensible to use triple glazing with an outer framed sealed unit window plus a secondary casement with 100 mm air gap.

For offices the double casement system is preferred and may even incorporate venetian blinds (remotely controlled) in the air gap. Although the gap should be as large as possible, it is often dictated by the thickness of the wall, e.g. brickwork is unlikely to provide the facility for more than a 150 mm gap. Curtain walling, however, can be made to provide 200 mm or even 300 mm air gaps and this is often used in medium-rise office buildings.

Structure-borne or impact sound may be considered as a separate problem (although many noise sources such as major road and rail traffic give rise to vibration through the ground – and thus the foundations – in addition to their airborne sound). Usually measures taken to reduce impact sound have little or no effect on airborne sound, the only area being 'coincident vibration', which may be reduced. One of the most frequent needs to insulate against impact sound is to reduce the noise of footsteps through upper floors. Carpeting, cork tiling or cushioned vinyl sheet will reduce the impact noise but 'floating floor' construction will also improve resistance to airborne sound. Four forms of floating floor are shown in fig. 8.20.

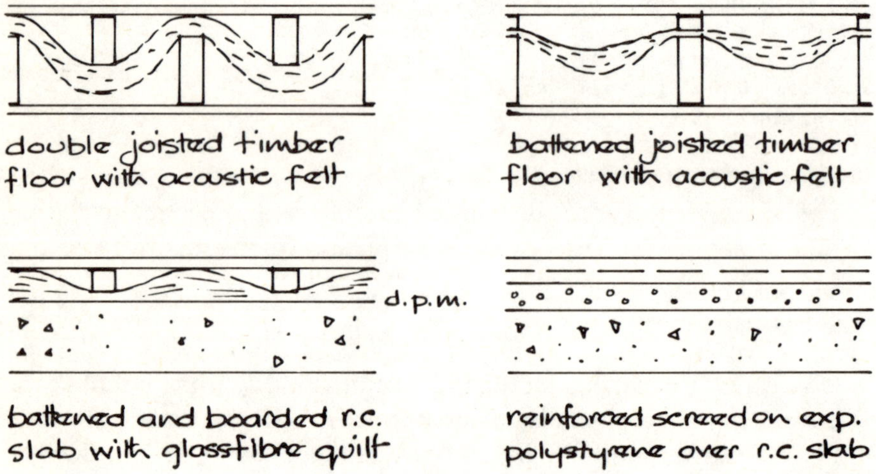

Fig 8.20 Floating floors – timber and r.c. slabs

When used in conjunction with reinforced concrete floor slabs, they may also provide other beneficial features such as improved thermal insulation (useful in flats) and a resilient surface for pedestrian movement. It should be noted that increasing the mass will not improve insulation from impact noise.

SOUND AND THE BUILDING REGULATIONS

Having examined the principles of sound insulation let us now consider the requirements of the Building Regulations in this respect. Party walls and floors, in conjunction with their associated structure, should have adequate resistance to airborne sound. Additionally, party floors should have adequate resistance to impact sound. There are two ways by which a party wall can be 'deemed to satisfy' these provisions.

(a) The wall may be built in accordance with the methods described in schedule 12 of the Regulations. Bonding with other walls will also influence sound transmission and due regard must be paid to this aspect.

(b) The sound insulation can be measured in the prescribed way in 'third-octave' bands. If the measured insulation at any band is less than that specified in the Regulations, it is counted as an 'adverse deviation'. The sum of all such deviations must not exceed 23 dB. It is important to realise that where an insulation is better than that specified at one frequency, it *cannot* be used to compensate for poor performance at other frequencies. Table 8.7 illustrates the recorded values for a wall subjected to this 'aggregate adverse deviation' procedure. Since the aggregate does not exceed 23dB, the wall satisfies the requirements of the Building Regulations. Similar criteria exist for party floors where the 'adverse deviation' concept can be applied additionally to impact insulation.

The standard for party walls and floors to housing are based on a series of grading curves devised by the Building Research Establishment (B.R.E.) and these are shown in fig. 8.21. The house party wall grading corresponds to the required values shown in table 8.7.

The overall performance of the structure is, of course, determined not only by the partitions but also by the way that these are fixed to other elements of the structure. Attention to detail of construction (and adequate supervision in the form of quality control) is therefore essential in providing good sound insulation. It is appropriate

Table 8.7 Sound reduction figures for a wall at various frequencies

Frequency (Hz)	Sound reduction required (dB)	Sound reduction measured (dB)	Adverse deviation (dB)
100	40	39	1
125	41	39	2
160	43	42	1
200	44	44	–
250	45	46	–
315	47	45	2
400	48	48	–
500	49	49	–
630	51	52	–
800	52	53	–
1000	53	53	–
1250	55	54	1
1600	56	55	1
2000	56	55	1
2500	56	56	–
3150	56	56	–
		Aggregate adverse deviation	9

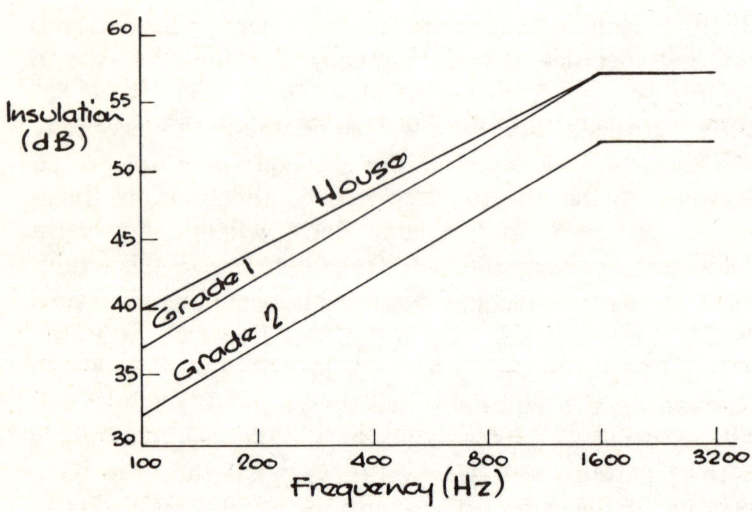

Fig 8.21 Airborne sound insulation grade curves

then to examine the features that will ensure that standards of insulation are met.

Walls

Solid walls such as party walls should have a mass of not less than 415 kg/m^2. For brickwork this effectively means a 'one-brick' wall whereas for aerated blockwork of normal density (not the extra-light thermal blocks) with 12 mm plaster each side, it would require a block thickness of 400 mm!

Cavity walls with a 50 mm cavity also require a mass of not less than 415 kg/m^2 but if the cavity is increased to 75 mm, using butterfly wall ties, it can be reduced to 250 kg/m^2 including plaster. The latter would allow an aerated block cavity wall comprising two 100 mm leaves, with 15 mm plaster to both outer faces, and a 75 mm cavity. Fortunately, the requirements do not apply to the cavity walls used externally, but using a half-brick outer skin, a 75 mm cavity (including 50 mm rigid thermal insulation), a 100 mm extra-lightweight aerated block inner skin and 12 mm plaster produces a value of 300 kg/m^2. The traditional 50 mm cavity would not meet such requirements!

Timber stud walls (timber-framed construction) have to use different techniques since the mass is very small. This involves using two separate studs, to provide an overall cavity between plasterboards of at least 250 mm, tied at intervals with light-gauge metal straps. Each stud should incorporate mineral wool quilting stapled all round to the studding (and horizontal fire stops) and the faces should be finished with *two* 12 mm sheets of plasterboard with joints staggered (this also reduces fire risks as well as breaking the air passage that might form).

Floors

The main purpose of the floating floor is to isolate the finish from the structural floor. The separating material should also continue around the perimeter and this can be 'masked' by the skirting board, 'dressing over' such packing. [*Note:* The skirtings must be fixed to the wall and *not* the floor in such situations.] Concrete slabs, whether precast or in situ, will not produce problems on 'mass' requirements but timber floors will probably need to include 'pugging', as mentioned earlier. They will probably incorporate some 'chicken wire' mesh and polythene sheet to support such pugging since it would be foolish to expect the plasterboard to support such material.

There should be two layers of plasterboard with staggered joints (as in the wall) and the floor finish must be tongued and grooved to prevent air passages. This *only* applies to party floors (which will also need to meet fire regulation requirements), *not* to floors within a dwelling, e.g. intermediate floors in maisonettes or houses.

Roofs

There are no statutory sound insulation requirements since they are an external element. However, there are many cases where sound insulation could be beneficial. Tiled roofing provides more mass than, say, ribbed aluminium but ventilation of the roof space, to minimise the risk of dry rot, will allow airborne sound to permeate the structure. In these cases it is advisable to locate the thermal insulation above the structural trusses (e.g. glass-fibre-backed roofing felt below the tiles) to produce a 'warm roof' and reduce the risk of condensation in the roof space. Alternatively, straw board and counter battens may be used below the sarking felt to minimise airborne sound. A flat roof is, of course, another answer to the problem but its advantages in terms of sound insulation have to be weighed against its disadvantages in terms of periodic maintenance.

CASE STUDY

A consideration of case study 6.1 from *Building and Civil Engineering Construction* Volume 1 poses some interesting problems in relation to sound control. Occupants of the upper floor of the office block may be subjected to noise from a variety of sources as follows:

(a) *Passing traffic:* assuming the building will be fully air conditioned, there is no need, even in summer, for windows to be opened (except for fire escape) and double casement windows should provide adequate sound insulation from airborne noise.

(b) *Roof-mounted plant:* Attention must be paid to possible structure-borne vibrations, and resilient mountings of dense synthetic rubber should be located under the machinery.

(c) *Occupants on other floors:* A floating floor will reduce both airborne sound (in both directions) and impact sound generated by the activities on the upper floor which might cause nuisance below.

(d) *Occupants on the same floor:* The use of a suspended ceiling will lead to many problems of 'flanking paths' between offices, and airborne transmission through the ducting of the air conditioning. Careful detail will need to be provided to minimise such passages. Direct transmission between rooms can be reduced by using twin leaf partitions but care must be taken that seals around the perimeter, and assembly joints, are adequate.

Exercises

8.1 (a) The velocity of sound is 340 m/s. What is the wavelength of a sound of frequency 200 Hz? (b) Given the density of steel as 7850 kg/m^3 and its modulus of elasticity as 207 kN/mm^2, what will be the velocity of sound through the metal?

8.2 (a) A sound is described as 60 dB above threshold. Express its intensity in watts per square metre (threshold intensity = 10^{-12} W/m^2). (b) Add 35 dB and 37 dB.

8.3 The noise analysis of a machine gives the following results:

Frequency (Hz)	125	250	500	1000	2000	4000
Level (dB)	55	70	71	73	75	71

Determine the total noise level.

8.4 (a) A noise of 85 dB is incident on a wall of insulation value 43 dB. What is the transmitted level? (b) What would the transmitted level be if the mass of the wall were doubled?

8.5 The sound insulation properties of a wall were measured at different frequencies and the results plotted against required decibel levels as shown in table 8.7. Calculate the A.A.D. for the wall and determine whether it meets the requirements of the Building Regulations with regard to sound insulation. The figures recorded are as follows:

Frequency (Hz)	100	125	160	200	250	315	400	500
Insulation (dB)	38	40	41	42	45	46	48	50
Frequency (Hz)	630	800	1000	1250	1600	2000	2500	3150
Insulation (dB)	51	51	52	54	55	55	56	56

Assignment

Sketch a floor plan and cross-section of a semi detached house near a busy main road and railway line. Draw the paths that might be taken by sound intruding into the house. Compare the performance of a 'traditional' house with that of a more lightweight construction and describe the precautions that the builder needs to take (for either form of building) to ensure a comfortably low noise level. How may the occupants improve on the builder's efforts?

PROJECT 1

A concrete estate road and provision for surface-level off-road parking is to be constructed to serve a small housing estate. The concrete is to be site mixed. It is expected that the maximum temperature range likely to be encountered after completion is 35°C. The materials are to be stored on site and all excavation, including topsoil, is to be carted away.

(a) The tip for excavated material is 20 km from the site, the route comprising 12 km of 50 km/h and 8 km of 80 km/h speed-restricted roads. Assuming the lorries can only travel at 70% of the speed when full that they achieve when empty, determine the minimum time cycle. Allow 20 min for discharging the load.

(b) Assuming the road to be cast in 5 m x 6 m bays with expansion joints to every third bay, what are the other joints? How wide should the expansion joints be? What material should be used? Detail both types of joint.

(c) What concrete mix should be specified for the road and parking areas? Describe two methods of ensuring the correct workability. Explain why the particular workability is specified. What would be the maximum size chosen for the aggregate? Explain how its grading is controlled. How quickly must the concrete be placed after mixing? Why is this important? What other materials are used in the construction, including curing?

(d) The groundworks include providing services to the houses and drainage from the houses and road/parking surfaces. Select suitable materials for all of these services, including inspection chambers, and explain the reasons for your selection.

(e) Illustrate the provision of temporary electricity to the site during construction and permanent supplies on its completion, explaining the reasons for using different voltages.

Approach the project from an analytical, science-based viewpoint, explaining the scientific logic behind the construction practice. Supplement your answers with extracts from digests, Codes of Practice, journals and laboratory tests.

PROJECT 2

The houses to the estate mentioned in project 1 are to be a mixture of terraced and detached. Select a constructional form, explaining the restricting influences of thermal and sound insulation on your selection.

(a) Produce a comprehensive list of materials used and explain why you have selected these in preference to others. Include a consideration of their performance in use as well as their ease of workability.

(b) Examine the risks of dampness occurring and explain how such risks must be overcome at the design stage. Indicate how quality control plays its part during the construction, including materials testing.

(c) Identify the structural elements of the buildings and sketch typical force and/or shearing force and bending moment diagrams for such elements.

(d) Select a suitable central heating system, including domestic water heating, and explain the principles of its circulation and heat output. Explain what is meant by 'efficiency' and illustrate how the overall system might be controlled.

(e) Describe the maintenance requirements for the fabric of the buildings, both external and internal, and show how the selection of decorative coatings and construction materials affects such maintenance.

Approach the project from a logical, science-based viewpoint, illustrating the scientific reasoning behind decisions on the selection of materials, components and techniques. Supplement your answers with extracts from digests, journals, site tests and technical reports.

Index